COMBINATORS

A CENTENNIAL VIEW

STEPHEN WOLFRAM

COMBINATORS

A CENTENNIAL VIEW

STEPHEN WOLFRAM

Combinators: A Centennial View

Copyright © 2021 Stephen Wolfram, LLC

Wolfram Media, Inc. | wolfram-media.com

ISBN-13: 978-1-57955-043-1 (hardback)
ISBN-13: 978-1-57955-044-8 (ebook)

Mathematics/Science

Cataloging-in-publication data available at:

Library of Congress Cataloging-in-Publication Data

Names: Wolfram, Stephen, author.
Title: Combinators: a centennial view / Stephen Wolfram.
Description: Champaign: Wolfram Media, Inc., [2021] | Includes
    bibliographical references and index.
Identifiers: LCCN 2021004419 (print) | LCCN 2021004420 (ebook) | ISBN
    9781579550431 (hardback) | ISBN 9781579550448 (ebook)
Subjects: LCSH: Combinatorial analysis. | Computer science – Mathematics. |
    Wolfram Language (computer program language) | AMS: Combinatorics –
    Historical. | Combinatorics – Explicit machine computation and programs
    (not the theory of computation or programming). | Computer science –
    Discrete mathematics in relation to computer science – Combinatorics in
    computer science. | Computer science – Discrete mathematics in relation
    to computer science – Graph theory (including graph drawing) in
    computer science.
Classification: LCC QA164.W65 2021 (print) | LCC QA164 (ebook) | DDC
    511/.60285–dc23
LC record available at https://lccn.loc.gov/2021004419
LC ebook record available at https://lccn.loc.gov/2021004420

For information about permission to reproduce selections from this book, contact permissions@wolfram.com. Sources for photos and archival materials that are not from the author's collection or in the public domain:

p. 167: Georg Olms Hildesheim; p. 168: Edizioni Cremonese; pp. 171, 173, 175–6, 178–9, 213, 215, 241–2, 244–7: *Mathematische Annalen*, Springer Nature; p. 184: *Monatshefte für Mathematik Physik*, Springer Nature; pp. 186–7: *Annals of Mathematics*, Johns Hopkins University Press; pp. 187, 268–71: *Am. J. of Mathematics*, Johns Hopkins University Press; p. 188: *J. of Symbolic Logic*, Cambridge University Press; p. 193: *Communications of the ACM*, Cambridge University Press; p. 195: Raymond Smullyan, Oxford U. Press; pp. 196, 272: North-Holland Pub. Co. (Cambridge University Press); pp. 213, 231–5, 240, 243, 254: Göttingen State and University Library; p. 217: Odessa State Archive, Ukraine; pp. 218, 264, 282: State Archives of Dnepropetrovsk Region, Ukraine; p. 219: ©Map Data 2015 Google; pp. 220–1: Dneiper City Encyclopedia, Ukraine; pp. 221–7: Odessa State Archive, Ukraine; p. 230: Cem Bozşahin, Stadtarchiv Göttingen; p. 238: Stadtarchiv Göttingen; p. 239: Stadtarchiv Göttingen (modified); p. 240: Goettinger-Tageblatt.de; pp. 248–252: Dr. Ludwig Bernays and ETH Zurich, Bernays Archive; p. 262: Deutsches Biographisches Archiv; p. 262: Staatsarchiv des Kantons Bern; p. 263: Semih Baskan, Ankara University; p. 265: Kim Heffernan; pp. 265–8: Haskell P. Curry papers, Pennsylvania State University; p. 283: Elena Zavoiskaia and Central State Archive of Moscow; p. 283: gwar.mil.ru; p. 284: pamyat-naroda.ru; p. 285: Daily Express/Reach Licensing.; p. 286: Osprey Publishing; pp. 287–8: world-war.ru/zabvenie-tradicii-baxrushinyx

Typeset with Wolfram Notebooks: wolfram.com/notebooks

Printed by Friesens, Manitoba, Canada. ∞ Acid-free paper. First edition. Revised second printing.

CONTENTS

Preface

I'd had it in my calendar for more than twenty years: December 7, 2020—the centenary of combinators. But how should it be celebrated? I'd first learned about combinators more than forty years ago—and a few times I'd explicitly used them. But they—and their origins—had always maintained a certain air of mystery for me.

So a few months ago I decided that the way I wanted to celebrate their centenary was finally to do my best to understand how combinators work, and how they came to be invented. This book is the result.

And on the actual day of December 7, 2020—exactly at the hour a hundred years after Moses Schönfinkel gave his talk about combinators—I was very pleased to be able to host an online event which brought together a large fraction of the living world experts on combinators.

Combinators—and their invention—are a tale of both triumph and tragedy. They defined a vision of computation that has a profound and timeless abstract elegance. But even a century later they continue to seem exotic and abstruse—and somehow fundamentally unsuitable for human consumption. In a different version of history the computer revolution might have been based on molecular rather than electronic computing, and combinators might have been at the center of it. But as it is, our computers are made of logic gates, and there aren't any combinators in sight.

But is that really true? As I worked on combinators in preparing for their centenary, I realized that combinators have had a much more profound effect on at least my thinking about computation than I'd ever imagined. For in the more than forty years that I've worked towards the full-scale computational language that is our modern Wolfram Language, the foundation of it all has been the idea of representing everything in terms of transformations of symbolic expressions. And I now realize that ultimately that idea—and everything that has flowed from it—can be traced back to combinators, and their introduction on December 7, 1920.

Combinators foreshadowed another big part of my life too. In trying to generalize the concept of mathematical models, I began many years ago to explore the computational universe of simple programs—and discovered that even extremely simple programs can generate immensely complex behavior, of the kind we see, for example in nature. Most of the programs I've studied over the years aren't specifically like combinators. But in preparing for the centenary of combinators I decided to apply the methods I've honed in exploring the computational universe to combinators—and discovered that many of the strongest phenomena I've found in the computational universe have been lurking largely unnoticed in combinators for a century.

When I first put the combinator centenary on my calendar twenty years ago—even though I didn't connect them as much with combinators as I do now—I already knew about symbolic expressions and computational language, and about what happens in the computational universe of simple programs. But—though I had some premonitions more than twenty years ago—it's only been in the last year and a half that I've come to realize with increasing certainty that our physical universe is fundamentally computational.

It's given me a new way to think about computation, deeply interwoven not only with physics but also with areas like metamathematics. And it's made me see that combinators—with some of their hard-to-understand features—fundamentally reflect some of how computation relates to physics and our experience of the physical world.

I'm so glad to have had the impetus from their centenary to get to know combinators a little better, and to be able to share what I've learned in this book. I think there's little doubt that the general development of computation has been the greatest intellectual achievement of the past century. But in that endeavor, combinators have been the road not taken. I hope here that I'm able to explore a little of the remarkable world of combinators, and what might lie ahead with them now that we begin to see what's possible.

Stephen Wolfram

April 30, 2021

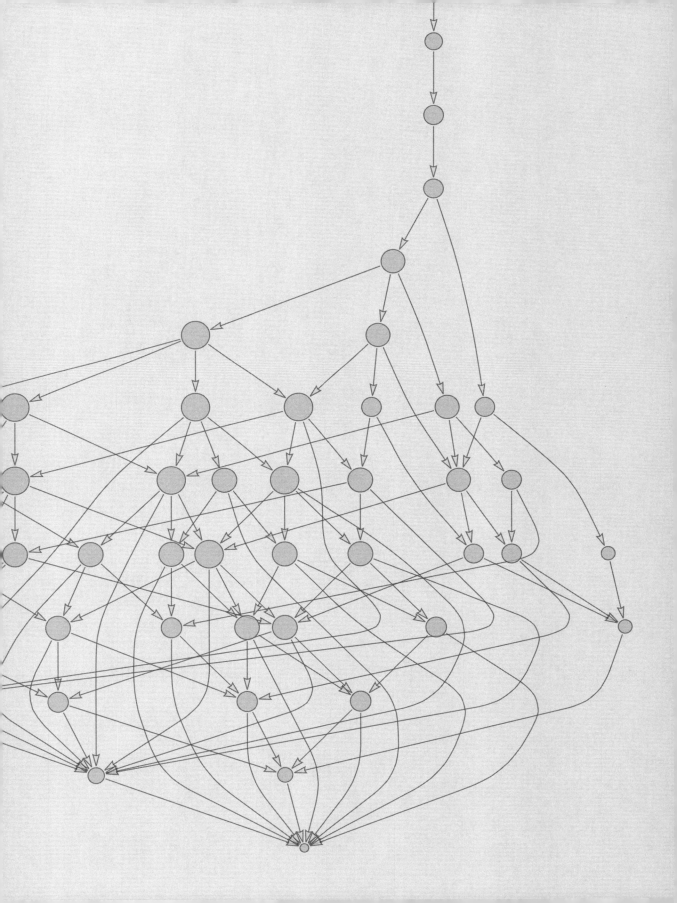

Combinators:
A Centennial View

Ultimate Symbolic Abstraction

Before Turing machines, before lambda calculus—even before Gödel's theorem—there were combinators. They were the very first abstract examples ever to be constructed of what we now know as universal computation—and they were first presented on December 7, 1920. In an alternative version of history our whole computing infrastructure might have been built on them. But as it is, for a century, they have remained for the most part a kind of curiosity—and a pinnacle of abstraction, and obscurity.

It's not hard to see why. In their original form from 1920, there were two basic combinators, s and k, which followed the simple replacement rules (now represented very cleanly in terms of patterns in the Wolfram Language):

s[x_][y_][z_] → x[z][y[z]]

k[x_][y_] → x

The idea was that any symbolic structure could be generated from some combination of s's and k's. As an example, consider a[b[a][c]]. We're not saying what a, b and c are; they're just symbolic objects. But given a, b and c how do we construct a[b[a][c]]? Well, we can do it with the s, k combinators.

Consider the (admittedly obscure) object

s[s[k[s]][s[k[k]][s[k[s]][k]]]][s[k[s[s[k][k]]]][k]]

(sometimes instead written **S(S(KS)(S(KK)(S(KS)K)))(S(K(S(SKK)))K)**).

Now treat this like a function and apply it to a, b, c s[s[k[s]][s[k[k]][s[k[s]][k]]]][s[k[s[s[k][k]]]][k]][a][b][c]. Then watch what happens when we repeatedly use the s, k combinator replacement rules:

```
s[s[k[s]][s[k[k]][s[k[s]][k]]]][s[k[s[s[k][k]]]][k]][a][b][c]
s[k[s]][s[k[k]][s[k[s]][k]]][a][s[k[s[s[k][k]]]][k]][a][b][c]
k[s][a][s[k[k]][s[k[s]][k]][a]][s[k[s[s[k][k]]]][k]][a][b][c]
s[s[k[k]][s[k[s]][k]][a]][s[k[s[s[k][k]]]][k][a]][b][c]
s[k[k]][s[k[s]][k]][a][b][s[k[s[s[k][k]]]][k][a][b]][c]
k[k][a][s[k[s]][k][a]][b][s[k[s[s[k][k]]]][k][a][b]][c]
k[s[k[s]][k][a]][b][s[k[s[s[k][k]]]][k][a][b]][c]
s[k[s]][k][a][s[k[s[s[k][k]]]][k][a][b]][c]
k[s][a][k[a]][s[k[s[s[k][k]]]][k][a][b]][c]
s[k[a]][s[k[s[s[k][k]]]][k][a][b]][c]
k[a][c][s[k[s[s[k][k]]]][k][a][b][c]]
a[s[k[s[s[k][k]]]][k][a][b][c]]
a[k[s[s[k][k]]][a][k[a]][b][c]]
a[s[s[k][k]][k[a]][b][c]]
a[s[k][k][b][k[a][b]][c]]
a[k[b][k[b]][k[a][b]][c]]
a[b[k[a][b]][c]]
a[b[a][c]]
```

Or, a tiny bit less obscurely:

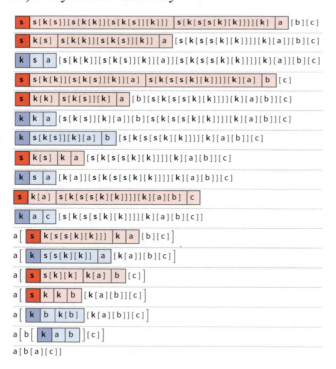

After a number of steps, we get a[b[a][c]]! And the point is that whatever symbolic construction we want, we can always set up some combination of s's and k's that will eventually do it for us—and ultimately be computation universal. They're equivalent to Turing machines, lambda calculus and all those other systems we know are universal. But they were discovered before any of these systems.

By the way, here's the Wolfram Language way to get the result above (//. repeatedly applies the rules until nothing changes anymore):

In[ ]:= s[s[k[s]][s[k[k]][s[k[s]][k]]]][s[k[s[s[k][k]]]][k]][a][b][c] //. {s[x_][y_][z_] → x[z][y[z]], k[x_][y_] → x}

Out[ ]= a[b[a][c]]

And, yes, it's no accident that it's extremely easy and natural to work with combinators in the Wolfram Language—because in fact combinators were part of the deep ancestry of the core design of the Wolfram Language.

For me, though, combinators also have another profound personal resonance. They're examples of very simple computational systems that turn out (as we'll see at length here) to show the same remarkable complexity of behavior that I've spent so many years studying across the computational universe.

A century ago—particularly without actual computers on which to do experiments—the conceptual framework that I've developed for thinking about the computational universe didn't exist. But I've always thought that of all systems, combinators were perhaps the earliest great "near miss" to what I've ended up discovering in the computational universe.

Computing with Combinators

Let's say we want to use combinators to do a computation on something. The first question is: how should we represent the "something"? Well, the obvious answer is: just use structures built out of combinators!

For example, let's say we want to represent integers. Here's an (at first bizarre-seeming) way to do that. Take s[k] and repeatedly apply s[s[k[s]][k]]. Then we'll get a sequence of combinator expressions:

```
s[k]
s[s[k[s]][k]][s[k]]
s[s[k[s]][k]][s[s[k[s]][k]][s[k]]]
s[s[k[s]][k]][s[s[k[s]][k]][s[s[k[s]][k]][s[k]]]]
s[s[k[s]][k]][s[s[k[s]][k]][s[s[k[s]][k]][s[s[k[s]][k]][s[k]]]]]
s[s[k[s]][k]][s[s[k[s]][k]][s[s[k[s]][k]][s[s[k[s]][k]][s[s[k[s]][k]][s[k]]]]]]
⋮
```

On their own, these expressions are inert under the s and k rules. But take each one (say *e*) and form *e*[s][k]. Here's what happens for example to the third case above when you then apply the s and k rules:

```
s[s[k[s]][k]][s[s[k[s]][k]][s[k]]][s][k]
s[k[s]][k][s[s[k[s]][k]][s[k]][s]][k]
k[s][s[k[s]][s[s[k[s]][k]][s[k]][s]]][k]
s[k[s]][s[s[k[s]][k]][s[k]][s]][k]
k[s][k][s[s[k[s]][k]][s[k]][s][k]]
s[s[s[k[s]][k]][s[k]][s][k]]
s[s[k[s]][k]][s][s[k[s]][s]][k]
s[k[s]][s][k[s]][s[k]][s]][k]
s[s[k[s]][s[k][s]][k]]
s[k[s][k][s[k][s][k]]]
s[s[s[k][s][k]]]
s[s[k[k][s[k]]]]
s[s[k]]
```

To get this in the Wolfram Language, we can use Nest, which nestedly applies functions:

In[◦]:= **Nest[f, x, 4]**

Out[◦]= f[f[f[f[x]]]]

Then the final result above is obtained as:

In[◦]:= **Nest[s[s[k[s]][k]], s[k], 2][s][k]** //. {s[x_][y_][z_]→x[z][y[z]], k[x_][y_]→x}

Out[◦]= s[s[k]]

Here's an example involving nesting 7 times:

In[ ]:= **Nest[s[s[k[s]][k]], s[k], 7][s][k]** *//.* {s[x_][y_][z_]→x[z][y[z]], k[x_][y_]→x}

Out[ ]= s[s[s[s[s[s[s[k]]]]]]]

So this gives us a (perhaps seemingly obscure) way to represent an integer *n*. Just form:

Nest[s[s[k[s]][k]], s[k], *n*]

This is a combinator representation of *n*, that we can "decode" by applying to [s][k]. OK, so given two integers represented this way, how would we add them together? Well, there's a combinator for that! And here it is:

s[k[s]][s[k[s[k[s]][k]]]]

If we call this plus, then let's compute plus[1][2][s][k], where 1 and 2 are represented by combinators:

```
s[k[s]][s[k[s[k[s]][k]]]][s[s[k[s]][k]][s[k]]][s[s[k[s]][k]][s[s[k[s]][k]][s[k]]]][s][k]
k[s][s[s[k[s]][k]][s[k]]][s[k[s[k[s]][k]]]][s[s[k[s]][k]][s[k]]]][s[s[k[s]][k]][s[s[k[s]][k]][s[k]]]][s][k]
s[s[k[s[k[s]][k]]]][s[s[k[s]][k]][s[k]]]][s[s[k[s]][k]][s[s[k[s]][k]][s[k]]]][s][k]
s[k[s[k[s]][k]]][s[s[k[s]][k]][s[k]]][s][s[s[k[s]][k]][s[s[k[s]][k]][s[k]]]][s][k]
k[s[k[s]][k]][s][s[s[k[s]][k]][s[k]]][s][s[s[k[s]][k]][s[s[k[s]][k]][s[k]]]][s][k]
s[k[s]][k][s[s[k[s]][k]][s[k]]][s][s[s[k[s]][k]][s[s[k[s]][k]][s[k]]]][s][k]
k[s][s[s[k[s]][k]][s[k]]][s][k[s[s[k[s]][k]][s[k]]][s]][s[s[k[s]][k]][s[s[k[s]][k]][s[k]]]][s][k]
s[k[s[s[k[s]][k]][s[k]]][s]]][s[s[k[s]][k]][s[s[k[s]][k]][s[k]]]][s][k]
k[s[s[k[s]][k]][s[k]]][s][k[s[s[k[s]][k]][s[k]]][s[s[k[s]][k]][s[k]]]][s][k]]
s[s[k[s]][k]][s[k]][s][s[s[k[s]][k]][s[s[k[s]][k]][s[k]]]][s][k]
s[k[s]][k][s][s[k[s]][s]][s[s[k[s]][k]][s[s[k[s]][k]][s[k]]]][s][k]
k[s][s][k[s]][s[k]][s][s[s[k[s]][k]][s[s[k[s]][k]][s[k]]]][s][k]
s[k[s]][s[k][s]][s[s[k[s]][k]][s[s[k[s]][k]][s[k]]]][s][k]
k[s][s[s[k[s]][k]][s[s[k[s]][k]][s[k]]]][s][k]][s[k][s][s[s[k[s]][k]][s[s[k[s]][k]][s[k]]]][s][k]]
s[s[k][s][s[s[k[s]][k]][s[s[k[s]][k]][s[k]]]][s][k]]]
s[k[s[s[k[s]][k]][s[k]]][s[s[k[s]][k]][s[k]]]][s][k]][s[s[s[k[s]][k]][s[s[k[s]][k]][s[k]]]][s][k]]]
s[s[s[k[s]][k]][s[s[k[s]][k]][s[k]]]][s][k]]
s[s[k[s]][k][s][s[s[k[s]][k]][s[k]]][s][k]]
s[k[s]][s][k[s]][s[s[k[s]][k]][s[k]]][s][k]]
s[s[k[s]][s[s[k[s]][k]][s[k]]][s][k]]
s[k[s]][k][s[s[k[s]][k]][s[k]][s][k]]
s[s[s[k[s]][k]][s[k]][s][k]]
s[s[k[s]][k][s][s[k][s][k]]]
s[s[k[s]][s][k[s]][s[k][s][k]]]
s[s[k[s]][s[k][s][k]]]
s[s[k[s]][k][s[k][s][k]]]
s[s[s[k][s][k]]]
s[s[s[k[k][s[k]]]]]
s[s[s[k]]]
```

It takes a while, but there's the result: 1 + 2 = 3.

Here's 4 + 3, giving the result s[s[s[s[s[s[s[k]]]]]]] (i.e. 7), albeit after 49 steps:

```
s[k[s]][s[s[s[k]][k]]]][s[s[k[s]][k]][[s[s[k]][k]][s[s[k[s]][k]][[s[s[k]][k]]]]]
...
s[s[s[s[s[s[k]]]]]]]
```

What about doing multiplication? There's a combinator for that too, and it's actually rather simple:

s[k[s]][k]

Here's the computation for 3 × 2—giving 6 after 58 steps:

```
s[k[s]][k][s[s[k[s]][k]][s[k[s]][k][...]]]
...
s[s[s[s[k]]]]]]
```

Here's a combinator for power:

s[k[s[s[k][k]]]][k]

And here's the computation of 3^2 using it (which takes 116 steps):

One might think this is a crazy way to compute things. But what's important is that it works, and, by the way, the basic idea for it was invented in 1920.

And while it might seem complicated, it's very elegant. All you need are s and k. Then you can construct everything from them: functions, data, whatever.

So far we're using what's essentially a unary representation of numbers. But we can set up combinators to handle binary numbers instead. Or, for example, we can set up combinators to do logic operations.

Imagine having k stand for true, and s[k] stand for false (so, like If[p,x,y], k[x][y] gives x while s[k][x][y] gives y). Then the minimal combinator for AND is just

s[s][k]

and we can check this works by computing a truth table (TT, TF, FT, FF):

s[s][k][k][k]	s[s][k][k][s[k]]	s[s][k][s[k]][k]	s[s][k][s[k]][s[k]]
s[k][k[k]][k]	s[k][k[k]][s[k]]	s[s][k]][k[s[k]]][k]	s[s][k]][k[s[k]]][s[k]]
k[k][k[k]][k]]	k[s[k]][k[k][s[k]]]	s[k][k][k[s[k]]][k]]	s[k][s[k]][k[s[k]]][s[k]]]
k	s[k]	k[k[s[k]][k]][k[k[s[k]][k]]]	k[k[s[k]][s[k]]][s[k][k[s[k]][s[k]]]]
		k[s[k]][k]	k[s[k]][s[k]]
		s[k]	s[k]

A search gives the minimal combinator expressions for the 16 possible 2-input Boolean functions:

15	▪▪▪▪	True	k[k[k]]	K(KK)
14	▪▪▪▫	Or	s[s[s]][s][s[k]]	S(SS)S(SK)
13	▪▪▫▪		s[k[s[s][k[k[k]]]]][k]	S(K(S(SS(K(KK)))))K
12	▪▪▫▫	First	k	K
11	▪▫▪▪	Implies	s[s][k[k[k]]]	SS(K(KK))
10	▪▫▪▫	Last	s[k]	SK
9	▪▫▫▪	Equal	s[s][k[s[s][k[k[k]]]][s]]	SS(K(S(SS(K(KK)))S))
8	▪▫▫▫	And	s[s][k]	SSK
7	▫▪▪▪	Nand	s[s[k[s[s][k[k[k]]]]]][s]	S(S(K(S(SS(K(KK))))))S
6	▫▪▪▫	Xor	s[s[s[s]][s[s[k]]][s]][k]	S(S(S(SS)(S(S(SK)))S))K
5	▫▪▫▪	Not	k[s[s][k[k[k]]][s]]	K(S(SS(K(KK)))S)
4	▫▪▫▫		s[k[s[s[s[k]]][s]]][k]	S(K(S(S(S(SK))S)))K
3	▫▫▪▪	Not	s[s][s[s[s[k]]][s]][k[k]]	SS(S(S(S(SK))S))(KK)
2	▫▫▪▫		s[s[s[k]]][s]	S(S(SK))S
1	▫▫▫▪	Nor	s[s[s[s]][k[k[k[k]]]]][k[s]]	S(S(S(SS(K(K(KK)))))(KS))
0	▫▫▫▫	False	k[k[s[k]]]	K(K(SK))

And by combining these (or even just copies of the one for NAND) one can make combinators that compute any possible Boolean function. And in fact in general one can—at least in principle—represent any computation by "compiling" it into combinators.

Here's a more elaborate example, from my book *A New Kind of Science*. This is a combinator that represents one step in the evolution of the rule 110 cellular automaton:

s[s[k[s]][s[k[s[s[k][k]]]][s[k[k]][s[s[s[s[s[k][k]][k[s[k]]]][
 k[s[s[k[s]][s[k[s[s[k][k]]]][s[k[k]][s[s[k[s]][s[k[s[s[k][k]]]][s[k[k]][s[s[k][k]][k[k]]]]]][
  s[k[k]][s[s[s[k][k]][k[s[k]]]][k[s[k]]]]]]]]]][s[k[k]][s[s[s[k][k]][k[s[k]]]][k[k]]]]]][
 s[k[s[s[s[k][k]][k[s[s[s[k][k]][k[s[s[s[k][k]][k[k]]][k[k]]]]][k[k]]]]]][s[k[k]][
 s[s[s[s[k][k]][k[s[k]]]][k[s[s[k[s]][s[k[s[s[k][k]][k[k]]][k[k]]]]][
  s[k[k]][s[s[k[s]][s[k[s[s[k][k]]]][s[k[k]][s[s[k][k]][k[s[k]]]]]]][s[k[k]][s[s[s[s[k][k]][k[k]]][
   k[s[k]]]]][s[s[s[s[k][k]][k[k]]][k[k]]][k[k]]][k[s[k]]]]]][s[s[s[s[s[k][k]][k[k]]][k[k]]][k[k]]][k[s[k]]]][s[k[k]]]]]][s[s[k[s]][
 s[k[s[s[k][k]]]][s[k[k]][s[s[k[s]][s[k[s[s[k][k]]]][s[k[k]][s[s[k][k]][k[k]]]]]][k[k[k]]]]]]][
 k[k[k]]]]][k[s[k]]]]]]]][k[k]]]]]][s[k[k]][s[k[s[s[k][s]][k]]]][s[s[k][k]][k[s[k]]]]]]

And, here from the book, are representations of repeatedly applying this combinator to compute—with great effort—three steps in the evolution of rule 110:

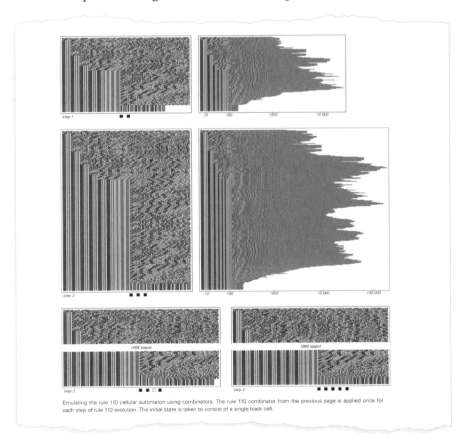

Emulating the rule 110 cellular automaton using combinators. The rule 110 combinator from the previous page is applied once for each step of rule 110 evolution. The initial state is taken to consist of a single black cell.

There's a little further to go, involving fixed-point combinators, etc. But basically, since we know that rule 110 is computation universal, this shows that combinators also are.

A Hundred Years Later...

Now that a century has passed, what should we think of combinators? In some sense, they still might be the purest way to represent computation that we know. But they're also very hard for us humans to understand.

Still, as computation and the computational paradigm advance, and become more familiar, it seems like on many fronts we're moving ever closer to core ideas of combinators. And indeed the foundational symbolic structure of the Wolfram Language—and much of what I've personally built over the past 40 years—can ultimately be seen as deeply informed by ideas that first arose in combinators.

Computation may be the single most powerful unifying intellectual concept ever. But the actual engineering development of computers and computing has tended to keep different aspects of it apart. There's data. There are data types. There's code. There are functions. There are variables. There's flow of control. And, yes, it may be convenient to keep these things apart in the traditional approach to the engineering of computer systems. But it doesn't need to be that way. And combinators show us that actually there's no need to have any of these distinctions: everything can be together, and can be made of the same, dynamic "computational stuff".

It's a very powerful idea. But in its raw form, it's also very disorienting for us humans. Because to understand things, we tend to rely on having "fixed anchors" to which we can attach meaning. And in pure, ever-changing seas of s, k combinators—like the ones we saw above—we just don't have these.

Still, there's a compromise—and in a sense that's exactly what's made it possible for me to build the full-scale computational language that the Wolfram Language now is. The point is that if we're going to be able to represent everything in the world computationally we need the kind of unity and flexibility that combinator-like constructs provide. But we don't just want raw, simple combinators. We need to in effect pre-define lots of combinator-like constructs that have particular meanings related to what we're representing in the world.

At a practical level, the crucial idea is to represent everything as a symbolic expression, and then to say that evaluating these expressions consists in repeatedly applying transformations to them. And, yes, symbolic expressions in the Wolfram Language are just like the expressions we've made out of combinators—except that instead of involving only s's and k's, they involve thousands of different symbolic constructs that we define to represent molecules, or cities or polynomials. But the key point is that—like with combinators—the things we're dealing with are always structurally just nested applications of pure symbolic objects.

Something we immediately learn from combinators is that "data" is really no different from "code"; they can both be represented as symbolic expressions. And both can be the raw material for computation. We also learn that "data" doesn't have to maintain any particular type or structure; not only its content, but also the way it is built up as a symbolic expression can be the dynamic output of a computation.

One might imagine that things like this would just be esoteric matters of principle. But what I've learned in building the Wolfram Language is that actually they're natural and crucially important in having convenient ways to capture computationally how we humans think about things, and the way the world is.

From the early days of practical computing, there was an immediate instinct to imagine that programs should be set up as sequences of instructions saying for example "take a thing, then do this to it, then do that" and so on. The result would be a "procedural" program like:

x = f[x]; x = g[x]; x = h[x]; x

But as the combinator approach suggests, there's a conceptually much simpler way to write this in which one's just successively applying functions, to make a "functional" program:

h[g[f[x]]]

(In the Wolfram Language, this can also be written h@g@f@x or x//f//g//h.)

Given the notion that everything is a symbolic expression, one's immediately led to have functions to operate on other functions, like

In[ ]:= **Nest[f, x, 6]**

Out[ ]= f[f[f[f[f[f[x]]]]]]

or:

In[ ]:= **ReverseApplied[f][a, b]**

Out[ ]= f[b, a]

This idea of such "higher-order functions" is quintessentially combinator informed —and very elegant and powerful. And as the years go by we're gradually managing to see how to make more and more aspects of it understandable and accessible in the Wolfram Language (think: Fold, MapThread, SubsetMap, FoldPair, ...).

OK, but there's one more thing combinators do—and it's their most famous: they allow one to set things up so that one never needs to define variables or name things. In typical programming one might write things like:

With[{x = 3}, 1+x^2]

f[x_] := 1+x^2

Function[x, 1+x^2]

x ⟼ 1+x^2

But in none of these cases does it matter what the actual name x is. The x is just a placeholder that's standing for something one's "passing around" in one's code.

But why can't one just "do the plumbing" of specifying how something should be passed around, without explicitly naming anything? In a sense a nested sequence of functions like f[g[x]] is doing a simple case of this; we're not giving a name to the result of g[x]; we're just feeding it as input to f in a "single pipe". And by setting up something like Function[x, 1+x^2] we're constructing a function that doesn't have a name, but which we can still apply to things:

In[◦]:= **Function[x, 1+x^2][4]**

Out[◦]= 17

The Wolfram Language gives us an easy way to get rid of the x here too:

In[◦]:= **(1+#^2) &[4]**

Out[◦]= 17

In a sense the # ("slot") here acts like a pronoun in a natural language: we're saying that whatever we're dealing with (which we're not going to name), we want to find "one plus the square of it".

OK, but so what about the general case? Well, that's what combinators provide a way to do.

Consider an expression like:

f[g[x][y]][y]

Imagine this was called q, and that we wanted q[x][y] to give f[g[x][y]][y]. Is there a way to define q without ever mentioning names of variables? Yes, here's how to do it with s, k combinators:

s[s[k[s]][s[k[s[k[f]]]][g]]][k[s[k][k]]]

There's no mention of x and y here; the combinator structure is just defining—without naming anything—how to "flow in" whatever one provides as "arguments". Let's watch it happen:

```
s[s[k[s]][s[k[s[k[f]]]][g]]][k[s[k][k]]][x][y]
s[k[s]][s[k[s[k[f]]]][g]][x][k[s[k][k]][x]][y]
k[s][x][s[k[s[k[f]]]][g][x]][k[s[k][k]][x]][y]
s[s[k[s[k[f]]]][g][x]][k[s[k][k]][x]][y]
s[k[s[k[f]]]][g][x][y][k[s[k][k]][x][y]]
k[s[k[f]]][x][g[x]][y][k[s[k][k]][x][y]]
s[k[f]][g[x]][y][k[s[k][k]][x][y]]
k[f][y][g[x][y]][k[s[k][k]][x][y]]
f[g[x][y]][k[s[k][k]][x][y]]
f[g[x][y]][s[k][k][y]]
f[g[x][y]][k[y][k[y]]]
f[g[x][y]][y]
```

Yes, it seems utterly obscure. And try as I might over the years to find a usefully human-understandable "packaging" of this that we could build into the Wolfram Language, I have so far failed.

But it's very interesting—and inspirational—that there's even in principle a way to avoid all named variables. Yes, it's often not a problem to use named variables in writing programs, and the names may even communicate useful information. But there are all sorts of tangles they can get one into.

It's particularly bad when a name is somehow global, and assigning a value to it affects (potentially insidiously) everything one's doing. But even if one keeps the scope of a name localized, there are still plenty of problems that can occur.

Consider for example:

$In[\circ]:=$ **Function[x, Function[y, 2 x + y]]**

It's two nested anonymous functions (AKA lambdas)—and here the x "gets" a, and y "gets" b:

$In[\circ]:=$ **Function[x, Function[y, 2 x + y]][a][b]**

$Out[\circ]:=$ 2 a + b

But what about this:

Function[x, Function[x, 2 x + x]]

The Wolfram Language conveniently colors things red to indicate that something bad is going on. We've got a clash of names, and we don't know "which x" is supposed to refer to what.

It's a pretty general problem; it happens even in natural language. If we write "Jane chased Bob. Jane ran fast." it's pretty clear what we're saying. But "Jane chased Jane. Jane ran fast." is already confused. In natural language, we avoid names with pronouns (which are basically the analog of # in the Wolfram Language). And because of the (traditional) gender setup in English "Jane chased Bob. She ran fast." happens to work. But "The cat chased the mouse. It ran fast." again doesn't.

But combinators solve all this, by in effect giving a symbolic procedure to describe what reference goes where. And, yes, by now computers can easily follow this (at least if they deal with symbolic expressions, like in the Wolfram Language). But the passage of a century—and even our experience with computation—doesn't seem to have made it much easier for us humans to follow it.

By the way, it's worth mentioning one more "famous" feature of combinators—that actually had been independently invented before combinators—and that these days, rather ahistorically, usually goes by the name "currying". It's pretty common—say in the Wolfram Language—to have functions that naturally take multiple arguments. GeoDistance[a, b] or Plus[a, b, c] (or a+b+c) are examples. But in trying to uniformize as much as possible, combinators just make all "functions" nominally have only one argument.

To set up things that "really" have multiple arguments, one uses structures like f[x][y][z]. From the point of standard mathematics, this is very weird: one expects "functions" to just "take an argument and return a result", and "map one space to another" (say real numbers to complex numbers).

But if one's thinking "sufficiently symbolically" it's fine. And in the Wolfram Language—with its fundamentally symbolic character (and distant ancestry in combinator concepts)—one can just as well make a definition like

f[x_][y_] := x + y

as:

f[x_, y_] := x + y

Back in 1980—even though I don't think I knew about combinators yet at that time—I actually tried in my SMP system that was a predecessor to the Wolfram Language the idea of having f[x][y] be able to be equivalent to f[x,y]. But it was a bit like forcing every verb to be intransitive—and there were many situations in which it was quite unnatural, and hard to understand.

Combinators in the Wild: Some Zoology

So far we've been talking about combinators that are set up to compute specific things that we want to compute. But what if we just pick possible combinators "from the wild", say at random? What will they do?

In the past, that might not have seemed like a question that was worth asking. But I've now spent decades studying the abstract computational universe of simple programs—and building a whole "new kind of science" around the discoveries I've made about how they behave. And with that conceptual framework it now becomes very interesting to look at combinators "in the wild" and see how they behave.

So let's begin at the beginning. The simplest s, k combinator expressions that won't just remain unchanged under the combinator rules have to have size 3. There are a total of 16 such expressions:

{ s[s][s] , s[s[s]] , s[s][k] , s[s[k]] , s[k][s] , s[k[s]] , s[k][k] ,
  s[k[k]] , k[s][s] , k[s[s]] , k[s][k] , k[s[k]] , k[k][s] , k[k[s]] , k[k][k] , k[k[k]]}

And none of them do anything interesting: they either don't change at all, or, as in for example k[s][s], they immediately give a single symbol (here s).

But what about larger combinator expressions? The total number of possible combinator expressions of size n grows like

{2, 4, 16, 80, 448, 2688, 16 896, 109 824, 732 160, 4 978 688}

or in general

$$2^n \text{ CatalanNumber}[n-1] = \frac{2^n \text{ Binomial}[2\,n-2, n-1]}{n}$$

or asymptotically:

$$\frac{8^n}{4 \sqrt{\pi}\; n^{3/2}}$$

At size 4, again nothing too interesting happens. With all the 80 possible expressions, the longest it takes to reach a fixed point is 3 steps, and that happens in 4 cases:

s[k][s][s]	s[k][s][k]	s[k][k][s]	s[k][k][k]
k[s][s[s]]	k[k][s[k]]	k[s][k[s]]	k[k][k[k]]
s	k	s	k

At size 5, the longest it takes to reach a fixed point is 4 steps, and that happens in 10 cases out of 448:

s[s][s][k][s]	s[s][s][k][k]	s[s][k][s][s]	s[s][k][s][s]	s[s][k][s][k]
s[k][s[k]][s]	s[k][s[k]][k]	s[s][k[s]][s]	s[k][s][s[s]]	s[s][k[s]][k]
k[s][s[k][s]]	k[k][s[k][k]]	s[s][k[s][s]]	k[s[s]][s[s[s]]]	s[k][k[s][k]]
s	k	s[s][s]	s[s]	s[k][s]

s[s][k]][s][k]	s[s][k][k][s]	s[s][k]][k][s]	s[s][k][k][k]	s[s][k]][k][k]
s[k][k][s[k]]	s[k][k[k]][s]	s[k][s][k[s]]	s[k][k[k]][k]	s[k][k][k[k]]
k[s][k]][k[s[k]]]	k[s][k[k]][s]	k[k[s]][s[k[s]]]	k[k][k[k][k]]	k[k[k]][k[k[k]]]
s[k]	s	k[s]	k	k[k]

At size 6, there is a slightly broader distribution of "halting times":

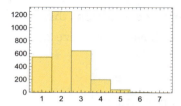

The longest halting time is 7, achieved by:

s[s[s]][s][k][s]	s[s[s]][s][k][k]
s[s][k][s[k]][s]	s[s][k][s[k]][k]
s[s[k]][k[s[k]]][s]	s[s[k]][k[s[k]]][k]
s[k][s][s[k]][s]]	s[k][k][k[s[k]][k]]
k[k[s[k]][s]][s[k[s[k]][s]]]	k[k[s[k]][k]][k[k[s[k]][k]]]
k[s[k]][s]	k[s[k]][k]
s[k]	s[k]

Meanwhile, the largest expressions created are of size 10 (in the sense that they contain a total of 10 s's or k's):

s[s[s]][s][s][s]	s[s][s][s[s]][s]	s[s[s]][s][s][k]	s[s][s][s[s]][k]
s[s][s][s[s]][s]	s[s[s]][s[s[s]]][s]	s[s][s][s[s]][k]	s[s[s]][s[s[s]]][k]
s[s[s]][s[s[s]]][s]	s[s][s][s[s]][s]]	s[s[s]][s[s[s]]][k]	s[s][k][s[s[s]][k]]
s[s][s][s[s[s]]][s]	s[s[s]][s]][s[s[s]]][s]]	s[k][s[s[s]][k]]	s[s[s]][k]][k[s[s[s]][k]]]
s[s[s]][s]][s[s[s]][s]]]		s[s[s]][k]][k[s[s[s]][k]]]	

The distribution of final sizes is a little odd:

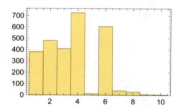

For size $n \le 5$, there's actually a gap with no final states of size $n-1$ generated. But at size 6, out of 2688 expressions, there are just 12 that give size 5 (about 0.4%).

OK, so what's going to happen if we go to size 7? Now there are 16,896 possible expressions. And there's something new: two never stabilize
(**S(SS)SSSS, SSS(SS)SS**):

{s[s[s]][s][s][s][s], s[s][s][s[s]][s][s]}

After one step, the first one of these evolves to the second, but then this is what happens over the next few steps (we'll see other visualizations of this later):

```
s[s][s][s[s]][s][s]
s[s[s]][s[s[s]]][s][s]
s[s][s][s[s[s]][s]][s]
s[s[s[s]][s]][s[s[s[s]][s]]][s]
s[s[s]][s][s][s[s[s[s]][s]][s]][s]]
s[s][s][s[s]][s[s[s[s]][s]][s]][s]]
s[s[s]][s[s[s]]][s[s[s[s]][s]][s]][s]]
s[s][s[s[s]][s]][s]][s[s[s]][s[s[s[s]][s]][s]]]
s[s[s[s]][s[s[s[s]][s]][s]][s]]][s[s[s[s]][s]][s]][s][s[s[s]][s[s[s[s]][s]][s]][s]]]]
```

The total size (i.e. LeafCount, or "number of s's") grows like:

{7, 8, 8, 11, 11, 11, 12, 17, 25, 33, 41, 50, 59, 87, 115, 149,

187, 215, 243, 272, 301, 389, 398, 413, 422, 431, 440, 450, 460, 491, 533}

A log plot shows that after an initial transient the size grows roughly exponentially:

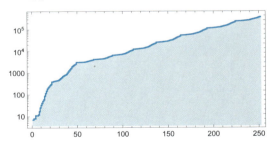

And looking at successive ratios one sees some elaborate fine structure:

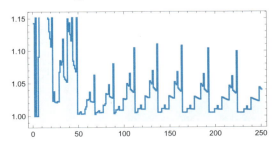

What is this ultimately doing? With a little effort, one finds that the sizes have a length-83 transient, followed by sequences of values of length $23 + 2n$, in which the second differences of successive sizes are given by:

Join[38 {0, 0, 0, 12, −17} 2^n +{0, 1, 0, −135, 189}, Table[0, n],
    38 {0, 1, 0, 0, 1, −1, 0, 0, 0, 4} 2^n +{12, −13, 0, 6, −7, 1, 0, 1, 0, −27},
    Table[0, n+2], 228 {0, 1, 0, 0, 1, −1} 2^n +2 {6, −20, 0, 3, −17, 14}]

The final sequence of sizes is obtained by concatenating these blocks and computing Accumulate[Accumulate[*list*]]—giving an asymptotic size that appears to be of the form $a \sqrt{t}\ 2^{b+\sqrt{t}}$. So, yes, we can ultimately "figure out what's going on" with this little size-7 combinator (and we'll see some more details later). But it's remarkable how complicated it is.

OK, but let's go back and look at the other size-7 expressions. The halting time distribution (ignoring the 2 cases that don't halt) basically falls off exponentially, but shows a couple of outliers:

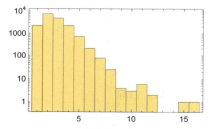

The maximum finite halting time is 16 steps, achieved by s[s[s[s]]][s][s][s] (**S(S(SS))SSS**):

```
s[s[s[s]]][s][s][s]
s[s[s]][s][s[s]][s]
s[s][s[s]][s[s[s]]][s]
s[s[s[s]]][s][s[s[s]]]][s]
s[s[s]][s][s[s][s[s[s]]]][s]]
s[s][s[s][s[s[s]]][s]][s[s][s[s[s]]][s]]]
s[s[s][s[s[s]]][s]]][s][s[s][s[s[s]]][s]][s[s][s[s[s]]][s]]]]
s[s[s][s[s[s]]][s]]][s][s[s][s[s[s]]][s]][s[s][s[s[s]]][s]]]]
s[s[s][s[s[s]]][s]]][s][s[s][s[s[s]]][s]][s[s][s[s[s]]][s]]]]
s[s[s][s[s[s]]][s]]][s[s[s][s[s[s]]][s]]][s[s][s][s[s][s[s[s]]][s]]]]]
s[s[s][s[s[s]]][s]]][s[s[s][s[s[s]]][s]]][s[s][s][s[s][s[s[s]]][s]]]]]
s[s[s][s[s[s]]][s]]][s[s[s][s[s[s]]][s]]][s[s][s[s][s[s[s]]][s]]][s[s[s][s[s[s]]][s]]]]]
s[s[s][s[s[s]]][s]]][s[s[s][s[s[s]]][s]]][s[s[s][s[s[s]]][s]]][s[s][s[s[s][s[s[s]]][s]]][s]]]]]
s[s[s][s[s[s]]][s]]][s[s[s][s[s[s]]][s]]][s[s[s][s[s[s]]][s]]][s[s[s][s[s[s]]][s]]][s[s[s][s[s[s]]][s]]]]]
s[s[s][s[s[s]]][s]]][s[s[s][s[s[s]]][s]]][s[s[s][s[s[s]]][s]]][s[s[s][s[s[s]]][s]]][s[s[s][s[s[s]]][s]]]]]
s[s[s][s[s[s]]][s]]][s[s[s][s[s[s]]][s]]][s[s[s][s[s[s]]][s]]][s[s[s][s[s[s]]][s]]][s[s[s][s[s[s]]][s]]]]]
```

And the distribution of final sizes is (with the maximum of 41 being achieved by the maximum-halting-time expression we've just seen):

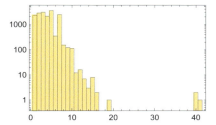

OK, so what happens at size 8? There are 109,824 possible combinator expressions. And it's fairly easy to find out that all but 76 of these go to fixed points within at most 50 steps (the longest survivor is s[s][s][s[s[s]]][k][k] (**SSS(S(SS))KK**), which halts after 44 steps):

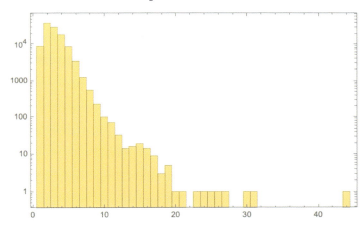

The final fixed points in these cases are mostly quite small; this is the distribution of their sizes:

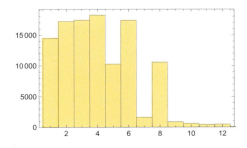 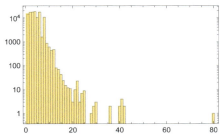

And here is a comparison between halting times and final sizes:

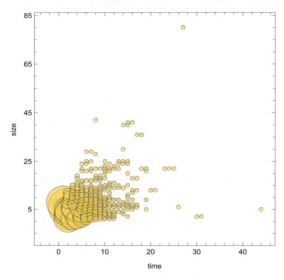

The outlier for size is s[s][k][s[s[s]][s]][s] (**SSK(S(SS)S)S**), which evolves in 27 steps to a fixed expression of size 80 (along the way reaching an intermediate expression of size 86):

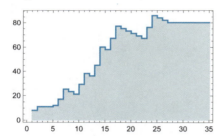

Among combinator expressions that halt in less than 50 steps, the maximum intermediate expression size of 275 is achieved for s[s][s][s[s[s][k]]][k] (**SSS(S(SSK))K**) (which ultimately evolves to s[s[s[s][k]]][k] after 26 steps):

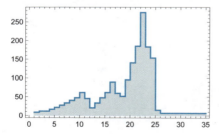

So what about size-8 expressions that don't halt after 50 steps? There are altogether 76—with 46 of these being inequivalent (in the sense that they don't quickly evolve to others in the set).

Here's how these 46 expressions grow (at least until they reach size 10,000):

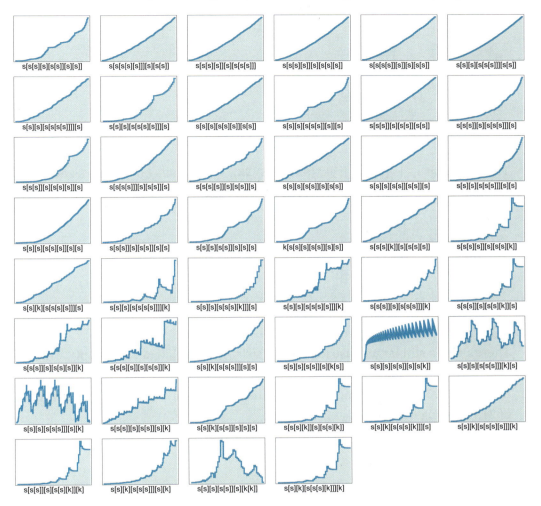

Some of these actually end up halting. In fact, s[s][s][s[s]][s][k[k]] (**SSS(SS)S(KK)**) halts after just 52 steps, with final result k[s[k][k[s[k][k]]]] (**K(SK(K(SKK))))**), having achieved a maximum expression size of 433:

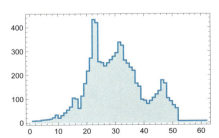

The next shortest halting time occurs for s[s][s][s[s[s]]][k][s] (**SSS(S(SS))KS**), which takes 89 steps to produce an expression of size 65:

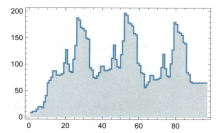

Then we have s[s][s][s[s[s]]][s][k] (**SSS(S(SS))SK**), which halts (giving the size-10 s[k][s[s[s[s[s]]][s]]][k]] (**SK(S(SS(S(SS))S))K**)), but only after 325 steps:

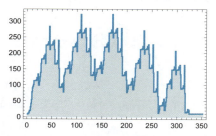

There's also a still-larger case to be seen: s[s[s][s]][s][s[s]][k] (**S(SSS)S(SS)K**), which exhibits an interesting "IntegerExponent-like" nested pattern of growth, but finally halts after 1958 steps, having achieved a maximum intermediate expression size of 21,720 along the way:

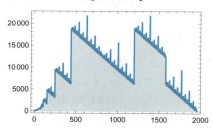

What about the other expressions? s[s][s][s[s]][s][s[k]] shows very regular $\sqrt{t}$ growth in size:

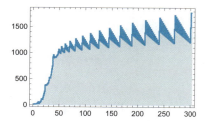

In the other cases, there's no such obvious regularity. But one can start to get a sense of what happens by plotting differences between sizes on successive steps:

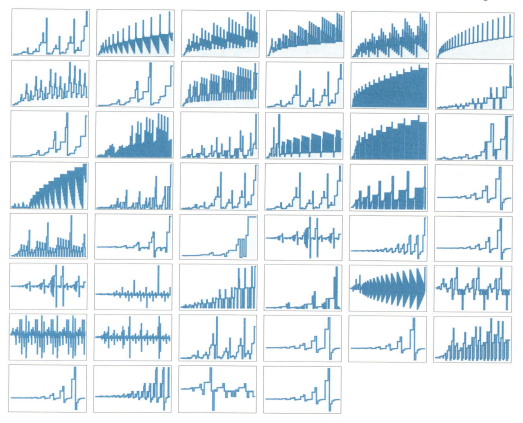

There are some obvious cases of regularity here. Several show a regular pattern of linearly increasing differences, implying overall t^2 growth in size:

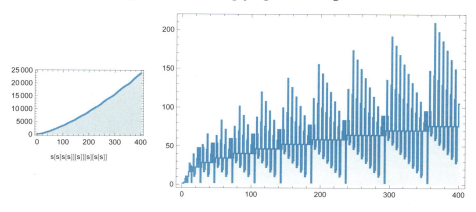

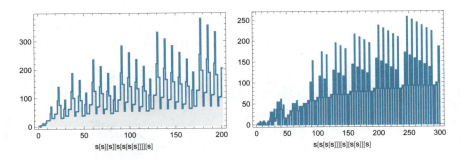

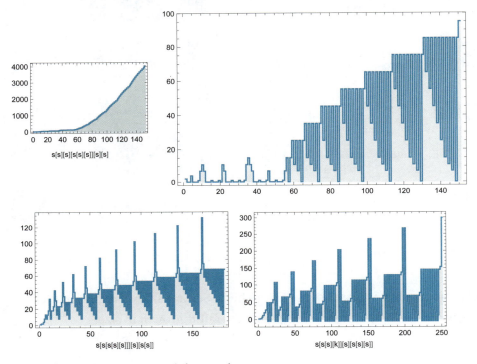

Others show regular $\sqrt{t}$ growth in differences, leading to $t^{3/2}$ growth in size:

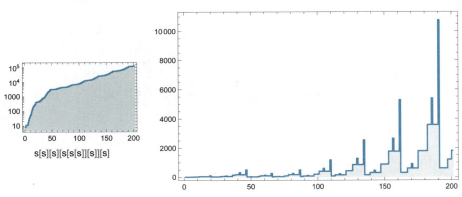

Others have pure exponential growth:

There are quite a few that have regular but below-exponential growth, much like the size-7 case s[s][s][s[s]][s][s] (**SSS(SS)SS**) with ~$2^{\sqrt{t}}$ growth:

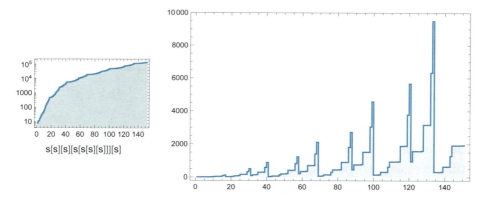

s[s][s][s[s][s]][s]

All the cases we've just looked at only involve s. When we allow k as well, there's for example s[s][s][s[s[s]]][k] (**SSS(S(SSS))K**)—which shows regular, essentially "stair-like" growth:

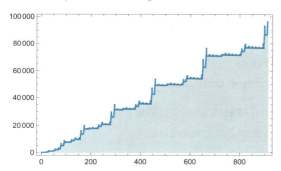

There's also a case like s[s[s]][s][s[s]][s][k] (**S(SS)S(SS)SK**):

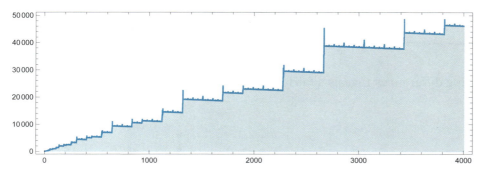

On a small scale, this appears somewhat regular, but the larger-scale structure, as revealed by taking differences, it doesn't seem so regular (though it does have a certain "IntegerExponent-like" look):

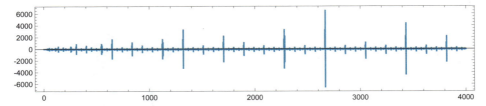

It's not clear what will happen in this case. The overall form of the behavior looks a bit similar to examples above that eventually terminate. Continuing for 50,000 steps, though, here's what's happened:

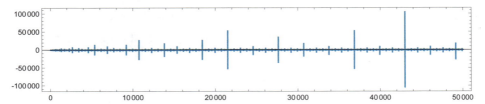

And in fact it turns out that the size-difference peaks continue to get higher—having values of the form $6 (17 \times 2^n + 1)$ and occurring at positions of the form $2 (9 \times 2^{n+2} + n - 18)$.

Here's another example: s[s][s][s[s]][s][k[s]] (**SSS(SS)S(KS)**). The overall growth in this case—at least for 200 steps—looks somewhat irregular:

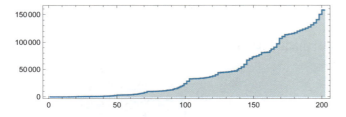

And taking differences reveals a fairly complex pattern of behavior:

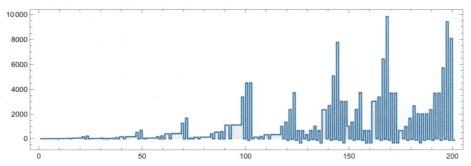

But after 1000 steps there appears to be some regularity to be seen:

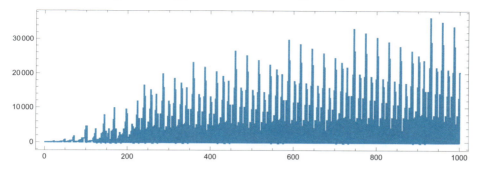

And even after 2000 steps the regularity is more obvious:

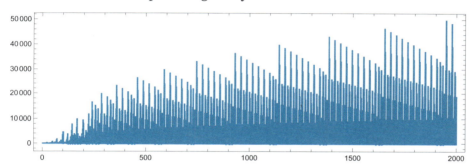

There's a long transient, but after that there are systematic peaks in the size difference, with the nth peak having height $16\,487 + 3320n$ and occurring at step $14n^2 + 59n + 284$. (And, yes, it's pretty weird to see all these strange numbers cropping up.)

What happens if we look at size-10 combinator expressions? There's a lot of repeating of behavior that we've seen with smaller expressions. But some new things do seem to happen.

After 1000 steps s[s][k][s[s][k][s[s]][s]][k] (**SSK(SSK(SS)S)K**) seems to be doing something quite complicated when one looks at its size differences:

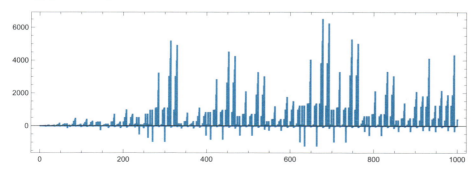

But it turns out that this is just a transient, and after 1000 steps or so, the system settles into a pattern of continual growth similar to ones we've seen before:

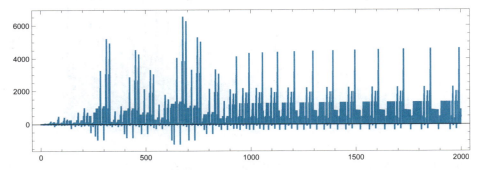

Another example is s[s][k][s[s][k][s[s]][s]][s] (**SSK(SSK(SS)S)S**). After 2000 steps there seems to be some regularity, and some irregularity:

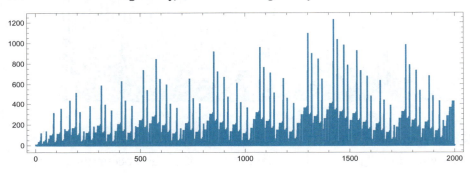

And basically this continues:

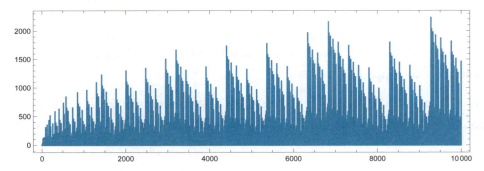

s[s][s][s[s[s[k]]]][s][s[k]] (**SSS(S(S(SK)))S(SK)**) is a fairly rare example of "nested-like" growth that continues forever (after a million steps, the size obtained is 597,871,806):

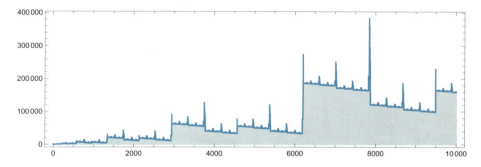

As a final example, consider s[s[s]][s][s][s][s[s][k[k]]] (**S**(**SS**)**SSS**(**SS**(**KK**))). Here's what this does for the first 1000 steps:

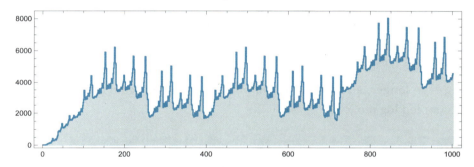

It looks somewhat complicated, but seems to be growing slowly. But then around step 4750 it suddenly jumps up, quickly reaching size 51,462:

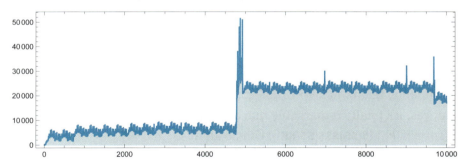

Keep going further, and there are more jumps:

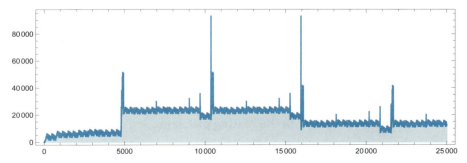

After 100,000 steps there's a definite pattern of jumps—but it's not quite regular:

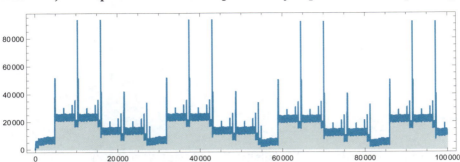

So what's going to happen? Mostly it seems to be maintaining a size of a few thousand or more. But then, after 218,703 steps, it dips down, to size 319. So, one might think, perhaps it's going to "die out". Keep going longer, and at step 34,339,093 it gets down to size 27, even though by step 36,536,622 it's at size 105,723.

Keep going even longer, and one sees it dipping down in size again (here shown in a downsampled log plot):

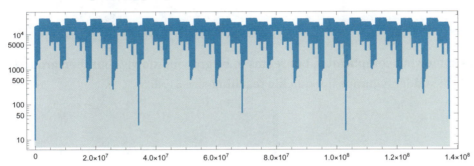

But, then, suddenly, boom. At step 137,356,329 it stops, reaching a fixed point of size 39. And, yes, it's totally remarkable that a tiny combinator expression like s[s[s]][s][s][s][s[s][k[k]]] (**S(SS)SSS(SS(KK))**) can do all this.

If one hasn't seen it before, this kind of complexity would be quite shocking. But after spending so long exploring the computational universe, I've become used to it. And now I just view each new case I see as yet more evidence for my Principle of Computational Equivalence.

A central fact about s, k combinators is that they're computation universal. And this tells us that whatever computation we want to do, it'll always be possible to "write a combinator program"—i.e. to create a combinator expression—that'll do it. And from this it follows that—just like with the halting problem for Turing machines—the problem of whether a combinator will halt is in general undecidable.

But the new thing we're seeing here is that it's difficult to figure out what will happen not just "in general" for complicated expressions set up to do particular computations but also for simple combinator expressions that one might "find in the wild". But the Principle of Computational Equivalence tells us why this happens.

Because it says that even simple programs—and simple combinator expressions—can lead to computations that are as sophisticated as anything. And this means that their behavior can be computationally irreducible, so that the only way to find out what will happen is essentially just to run each step and see what happens. So then if one wants to know what will happen in an infinite time, one may have to do an effectively infinite amount of computation to find out.

Might there be another way to formulate our questions about the behavior of combinators? Ultimately we could use any computation universal system to represent what combinators do. But some formulations may connect more immediately with existing ideas—say mathematical ones. And for example I think it's conceivable that the sequences of combinator sizes we've seen above could be obtained in a more "direct numerical way", perhaps from something like nestedly recursive functions (I discovered this particular example in 2003):

f[n_] := 3 f[n − f[n − 1]]

f[n_ /; n < 1] = 1

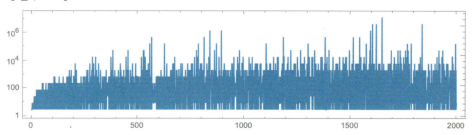

Visualizing Combinators

One of the issues in studying combinators is that it's so hard to visualize what they're doing. It's not like with cellular automata where one can make arrays of black and white cells and readily use our visual system to get an impression of what's going on. Consider for example the combinator evolution:

```
s[s][k][s[s[s]][s]][s]
s[s[s[s]][s]][k[s[s[s]][s]]][s]
s[s[s]][s][s][k[s[s[s]][s]][s]][s]
s[s][s[s[s]][k[s[s[s]][s]][s]][s]
s[s[s]][s[s[s]]][k[s[s[s]][s]][s]]
s[s][k[s[s[s]][s]][s]][s]][s[s[s]][k[s[s[s]][s]][s]]]
s[s[s[s]][k[s[s[s]][s]][s]][s]][k[s[s[s]][s]][s]][s][s[s[s]][k[s[s[s]][s]][s]]]]
s[s[s[s]][s[s[s]][s]]][k[s[s[s]][s]][s]][s][s[s[s]][k[s[s[s]][s]][s]]]]
```

In a cellular automaton the rule would be operating on neighboring elements, and so there'd be locality to everything that's happening. But here the combinator rules are effectively moving whole chunks around at a time, so it's really hard to visually trace what's happening.

But even before we get to this issue, can we make the mass of brackets and letters in something like

Out[ ]= s[s[s[s]][k[s[s[s]][s]][s]][s]][k[s[s[s]][s]][s]][s][s[s[s]][k[s[s[s]][s]][s]]]]

easier to read? For example, do we really need all those brackets? In the Wolfram Language, for example, instead of writing

a[b[c[d[e]]]]

we can equivalently write

In[ ]:= **a@b@c@d@e**

thereby avoiding brackets.

But using @ doesn't avoid all grouping indications. For example, to represent

a[b][c][d][e]

with @ we'd have to write:

In[ ]:= **(((a@b)@c)@d)@e**

In our combinator expression above, we had 24 pairs of brackets. By using @, we can reduce this to 10:

(s @ (s @ s @ s) @ (k @ (s @ s @ s) @ s) @ s) @ ((k @ (s @ s @ s) @ s) @ s) @ (s @ s @ s) @ (k @ (s @ s @ s) @ s) @ s

And we don't really need to show the @, so we can make this smaller:

(s(sss)(k(sss)s)s)((k(sss)s)s)(sss)(k(sss)s)s

When combinators were first introduced a century ago, the focus was on "multi-argument-function-like" expressions such as a[b][c] (as appear in the rules for s and k), rather than on "nested-function-like" expressions such as a[b[c]]. So instead of thinking of function application as "right associative"—so that a[b[c]] can be written without parentheses as a @ b @ c—people instead thought of function application as left associative—so that a[b][c] could be written without parentheses. (Confusingly, people often used @ as the symbol for this left-associative function application.)

As it's turned out, the f[g[x]] form is much more common in practice than f[g][x], and in 30+ years there hasn't been much of a call for a notation for left-associative function application in the Wolfram Language. But in celebration of the centenary of combinators, we've decided to introduce Application (indicated by •) to represent left-associative function application.

So this means that a[b][c][d][e] can now be written

a•b•c•d•e

without parentheses. Of course, now a[b[c[d[e]]]] needs parentheses:

a•(b•(c•(d•e)))

In this notation the rules for s and k can be written without brackets as:

{s•x_•y_•z_ → x•z•(y•z), k•x_•y_ → x}

Our combinator expression above becomes

s•(s•(s•s)•(k•(s•(s•s)•s)•s))•(k•(s•(s•s)•s)•s•(s•(s•s)•(k•(s•(s•s)•s)•s)))

or without the function application character

s(s(ss)(k(s(ss)s)s))(k(s(ss)s)s(s(ss)(k(s(ss)s)s)))

which now involves 13 pairs of parentheses.

Needless to say, if you consider all possible combinator expressions, left and right associativity on average do just the same in terms of parenthesis counts: for size-n combinator expressions, both on average need $(n-2)/2$ pairs; the number of cases needing k pairs is

$$\text{Binomial}[n-1, k-1] \; \frac{\text{Binomial}[n, k-1]}{k}$$

(the "Catalan triangle"). (Without associativity, we're dealing with our standard representation of combinator expressions, which always requires $n-1$ pairs of brackets.)

By the way, the number of "right-associative" parenthesis pairs is just the number of subparts of the combinator expression that match _[_][_], while for left-associative parenthesis pairs it's the number that match _[_[_]]. (The number of brackets in the no-associativity case is the number of matches of _[_].)

If we look at the parenthesis/bracket count in the evolution of the smallest non-terminating combinator expression from above s[s][s][s[s]][s][s] (otherwise known as s•s•s•(s•s)•s•s) we find:

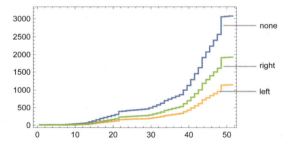

Or in other words, in this case, left associativity leads on average to about 62% of the number of parentheses of right associativity. We'll look at this in more detail later, but for growing combinator expressions, it'll almost always turn out to be the case that left associativity is the "parenthesis-avoidance winner".

But even with our "best parenthesis avoidance" it's still very hard to see what's going on from the textual form:

left	right	none
sss(ss)ss	((((ss)s)ss)s)s	s[s][s][s[s]][s][s]
s(ss)(s(ss))ss	(((sss)sss)s)s	s[s[s]][s[s[s]]][s][s]
sss(s(ss)s)s	(((ss)s)(sss)s)s	s[s][s][s[s]][s]][s]
s(s(ss)s)(s(s(ss)s))s	((s(sss)s)s(sss)s)s	s[s[s[s]]][s]][s[s[s[s]][s]]][s]
s(ss)ss(s(s(ss)s)s)	(((sss)s)s)(s(sss)s)s	s[s[s]][s][s][s[s[s[s]][s]][s]]
sss(ss)(s(s(ss)s)s)	(((ss)s)ss)(s(sss)s)s	s[s][s][s[s]][s[s[s[s]][s]][s]]
s(ss)(s(ss))(s(s(ss)s)s)	((sss)sss)(s(sss)s)s	s[s[s]][s[s[s]]][s[s[s[s]][s]][s]]
ss(s(s(ss)s)s)(s(ss)(s(s(ss)s)s))	((ss)(s(sss)s)s)(sss)(s(sss)s)s)	s[s][s[s[s[s]][s]][s]][s]][s[s[s]][s[s[s[s]][s]][s]]][s]]

So what about getting rid of parentheses altogether? Well, we can always use so-called Polish (or Łukasiewicz) "prefix" notation—in which we write f[x] as •fx and f[g[x]] as •f•gx. And in this case our combinator expression from above becomes:

••s••s•ss••k••s•ssss•••k••s•ssss••s•ss••k••s•ssss

Alternatively—like a traditional HP calculator—we can use reverse Polish "postfix" notation, in which f[x] is fx• and f[g[x]] is fgx•• (and • is like HP ENTER):

ssss••ksss••s••s•••ksss••s••s•sss••ksss••s••s••••

The total number of • symbols is always equal to the number of pairs of brackets in our standard "non-associative" functional form:

•••••sss•ssss	ss•s•ss••s•s•
••••s•ss•s•ssss	sss••sss•••s•s•
••••sss••s•ssss	ss•s•sss••s••s•
•••s••s•sss•s••s•ssss	ssss••s••ssss••s•••s•
••••s•ssss••s••s•ssss	sss••s•s•ssss••s••s••
••••sss•ss••s••s•ssss	ss•s•ss••ssss••s••s••
•••s•ss•s•ss••s••s•ssss	sss••sss•••ssss••s••s••
•••ss••s••s•ssss•••s•ss••s••s•ssss	ss•ssss••s••s•sss••ssss••s••s•••
Polish	*reverse Polish*

What if we look at this on a larger scale, "cellular automaton style", with s being ■ and • being ▢? Here's the not-very-enlightening result:

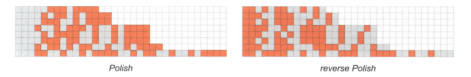

Polish   reverse Polish

Running for 50 steps, and fixing the aspect ratio, we get (for the Polish case):

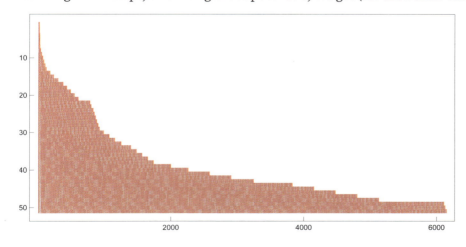

We can make the same kinds of pictures from our bracket representation too. We just take a string like s[s][s][s[s]][s][s] and render each successive character as a cell of some color. (It's particularly easy if we've only got one basic combinator—say s—because then we only need colors for the opening and closing brackets.) We can also make "cellular automaton–style" pictures from parenthesis representations like **SSS(SS)SS**. Again, all we do is render each successive character as a cell of some color.

The results essentially always tend to look much like the reverse Polish case above. Occasionally, though, they reveal at least something about the "innards of the computation". Like here's the terminating combinator expression s[s][s][s[s[s]]][k][s] from above, rendered in right-associative form:

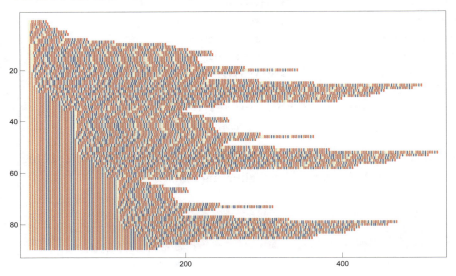

Pictures like this in a sense convert all combinator expressions to sequences. But combinator expressions are in fact hierarchical structures, formed by nested invocations of symbolic "functions". One way to represent the hierarchical structure of

s[s][s][s[s]][s][s]

is through a hierarchy of nested boxes:

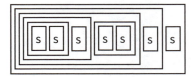

We can color each box by its depth in the expression:

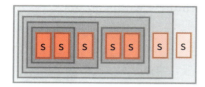

But now to represent the expression all we really need to do is show the basic combinators in a color representing its depth. And doing this, we can visualize the terminating combinator evolution above as:

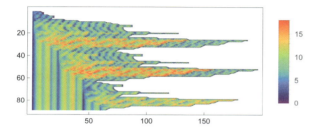

We can also render this in 3D (with the height being the "depth" in the expression):

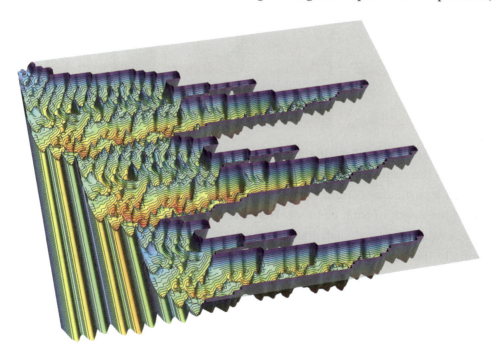

To test out visualizations like these, let's look (as above) at all the size-8 combinator expressions with distinct evolutions that don't terminate within 50 steps. Here's the "depth map" for each case:

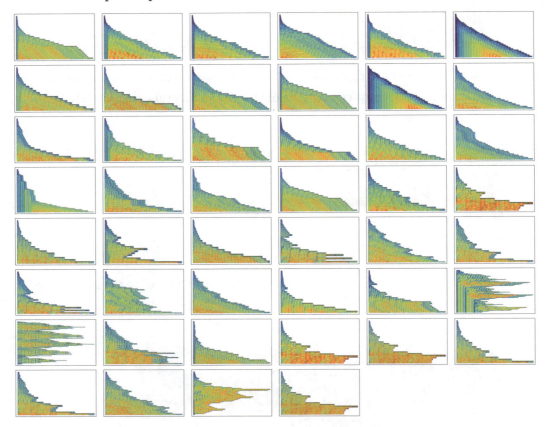

In these pictures we're drawing a cell for each element in the "stringified version" of the combinator expression at each step, then coloring it by depth. But given a particular combinator expression, one can consider other ways to indicate the depth of each element. Here are a few possibilities, shown for step 8 in the evolution of s[s][s][s[s]][s][s] (**SSS(SS)SS**) (note that the first of these is essentially the "indentation level" that might be used if each s, k were "pretty printed" on a separate line):

s[s[s[s]][s[s[s]][s]][s]]][s[s[s[s]][s]][s][s[s[s]][s[s[s]][s]][s]]]]

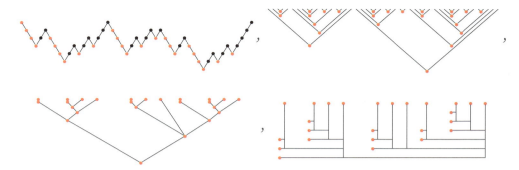

And this is what one gets on a series of steps:

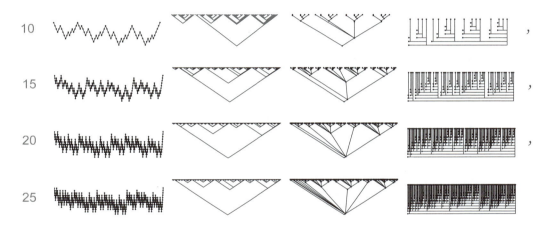

But in a sense a more direct visualization of combinator expressions is as trees, as for example in:

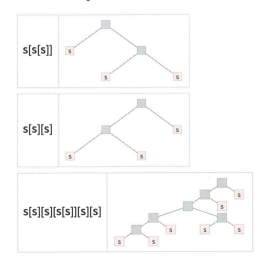

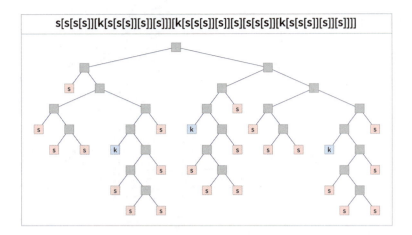

s[s[s[s]][k[s[s[s]][s]][s]]]][k[s[s[s]][s]][s]][s[s[s]][k[s[s[s]][s]][s]]]]

Note that these trees can be somewhat simplified by treating them as left or right "associative", and essentially pulling left or right leaves into the "branch nodes".

But using the original trees, we can ask for example what the trees for the expressions produced by the evolution of s[s][s][s[s]][s][s] (**SSS(SS)SS**) are. Here are the results for the first 15 steps:

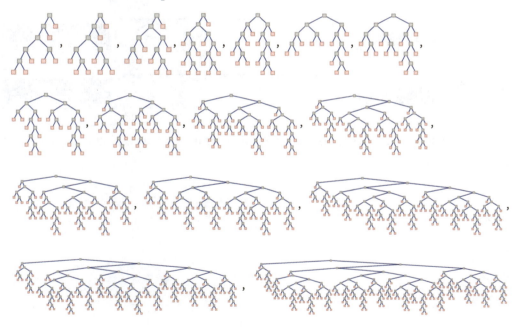

In a different rendering, these become:

OK, so these are representations of the combinator expressions on successive steps. But where are the rules being applied at each step? As we'll discuss in much more detail in the next section, in the way we've done things so far we're always doing just one update at each step. Here's an example of where the updates are happening in a particular case:

s	s	s	s[s[s]]	[k][s]

| s | s[s[s]] | s[s[s[s]]] | k | [s] |

| s | s[s] | k | s[s[s[s]]][k] | [s] |

| s | s | s[s[s[s]]][k] | k[s[s[s[s]]][k]] | [s] |

| s | k[s[s[s[s]]][k]] | s[s[s[s]]][k][k[s[s[s[s]]][k]]] | s |

| k | s[s[s[s]]][k] | s | [s[s[s[s]]][k][k[s[s[s[s]]][k]]][s]] |

| s | s[s[s]] | k | s[s[s[s]]][k][k[s[s[s[s]]][k]]][s] |

| s | s[s] | s[s[s[s]]][k][k[s[s[s[s]]][k]]][s] | k[s[s[s[s]]][k][k[s[s[s[s]]][k]]][s]] |

Continuing longer we get (note that some lines have wrapped in this display):

A feature of the way we're writing out combinator expressions is that the "input" to any combinator rule always corresponds to a contiguous span within the expression as we display it. So when we show the total size of combinator expressions on each step in an evolution, we can display which part is getting rewritten:

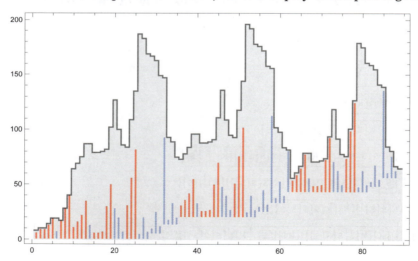

Notice that, as expected, application of the S rule tends to increase size, while the K rule decreases it.

Here is the distribution of rule applications for all the examples we showed above:

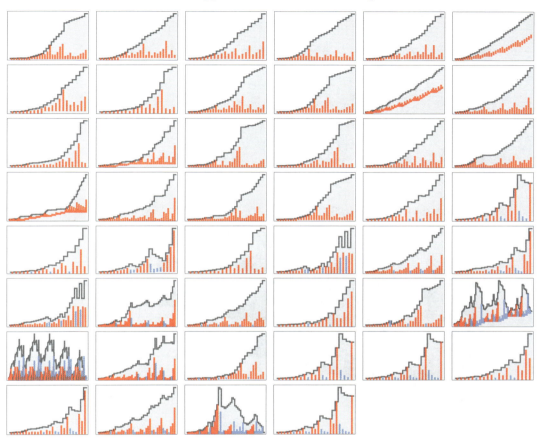

We can combine multiple forms of visualization by including depths:

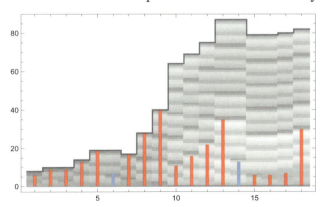

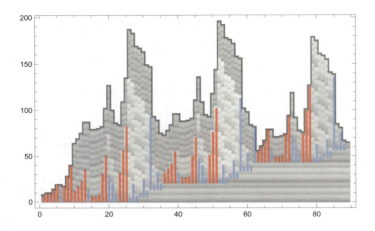

We can also do the same in 3D:

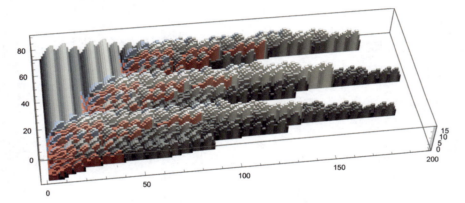

So what about the underlying trees? Here are the S, K combinator rules in terms of trees:

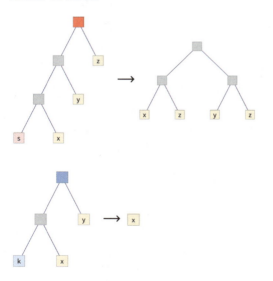

And here are the updates for the first few steps of the evolution of
s[s][s][s[s[s]]][k][s] (**SSS(S(SS))KS**):

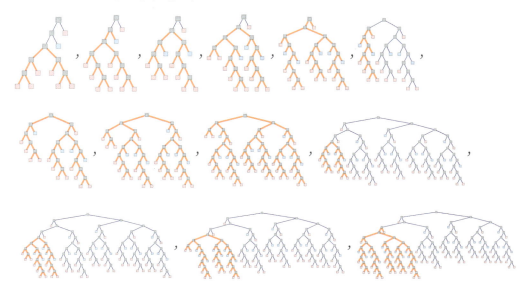

In these pictures we are effectively at each step highlighting the "first" subtree
matching s[_][_][_] or k[_][_]. To get a sense of the whole evolution, we can
also simply count the number of subtrees with a given general structure (like
_ [_][_] or _ [_ [_]]) that occur at a given step (see also below):

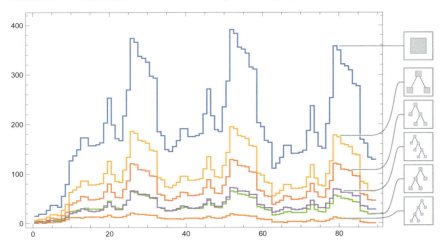

One more indication of the behavior of combinators comes from looking at tree
depths. In addition to the total depth (i.e. Wolfram Language Depth) of the combinator tree, one can also look at the depth at which update events happen (here
with the total size shown underneath):

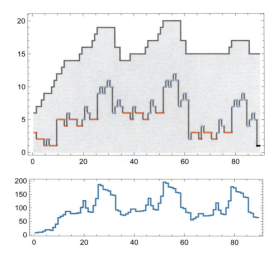

Here are the depth profiles for the rules shown above:

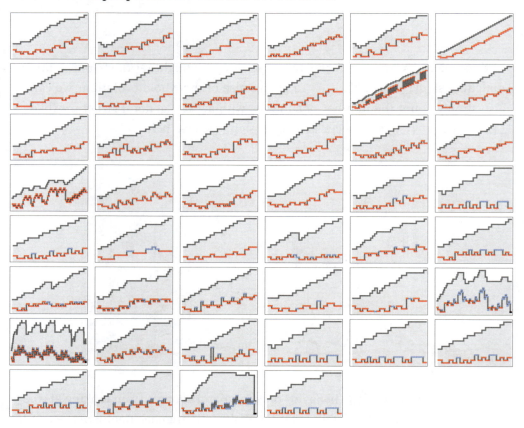

Not surprisingly, total depth tends to increase when growth continues. But it is notable that—except when termination is near at hand—it seems like (at least with our current updating scheme) updates tend to be made to "high-level" (i.e. low-depth) parts of the expression tree.

When we write out a combinator expression like the size-33

s[s[s[s]][s[s[s[s]][s]][s]]][s[s[s]][s][s[s[s]][s[s[s[s]][s]][s]]][s[s[s]][s[s[s[s]][s]][s]]]]

or show it as a tree

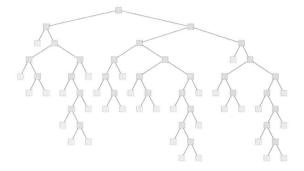

we're in a sense being very wasteful, because we're repeating the same subexpressions many times. In fact, in this particular expression, there are 65 subexpressions altogether—but only 11 distinct ones.

So how can we represent a combinator expression making as much use as possible of the commonality of these subexpressions? Well, instead of using a tree for the combinator expression, we can use a directed acyclic graph (DAG) in which we start from a node representing the whole expression, and then show how it breaks down into shared subexpressions, with each shared subexpression represented by a node.

To see how this works, let's consider first the trivial case of f[x]. We can represent this as a tree—in which the root represents the whole expression f[x], and has one connection to the head f, and another to the argument x:

The expression f[g[x]] is still a tree:

But in f[f[x]] there is a "shared subexpression" (which in this case is just f), and the graph is no longer a tree:

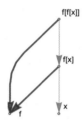

For f[x][f[x][f[x]]], f[x] is a shared subexpression:

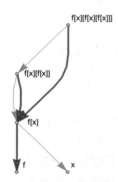

For s[s][s][s[s]][s][s] things get a bit more complicated:

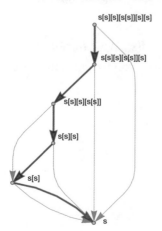

For the size-33 expression above, the DAG representation is

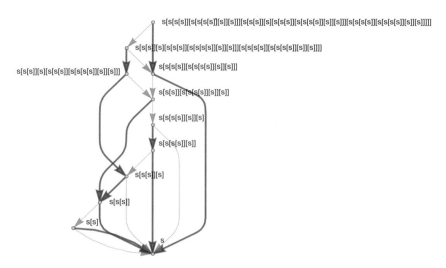

where the nodes correspond to the 11 distinct subexpressions of the whole expression that appears at the root.

So what does combinator evolution look like in terms of DAGs? Here are the first 15 steps in the evolution of s[s][s][s[s]][s][s]:

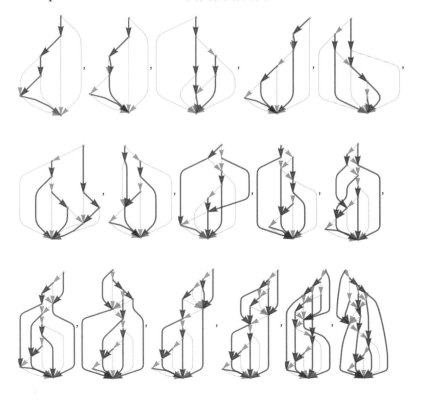

And here are some later steps:

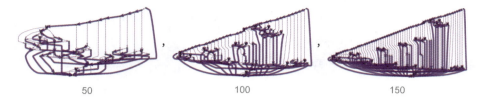

Sharing all common subexpressions is in a sense a maximally reduced way to specify a combinator expression. And even when the total size of the expressions is growing roughly exponentially, the number of distinct subexpressions may grow only linearly—here roughly like $1.24\,t$:

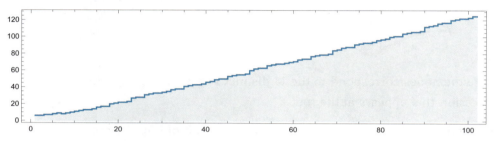

Looking at successive differences suggests a fairly simple pattern:

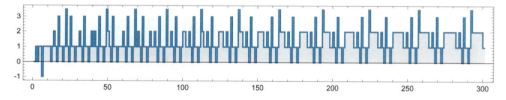

Here are the DAG representations of the result of 50 steps in the evolution of the 46 "growing size-7" combinator expressions above:

It's notable that some of these show considerable complexity, while others have a rather simple structure.

Updating Schemes and Multiway Systems

The world of combinators as we've discussed it so far may seem complicated. But we've actually so far been consistently making a big simplification. And it has to do with how the combinator rules are applied.

Consider the combinator expression:

s[s[s][s][s][s][k[k][s]][s]]][s][s][s[k[s][k]][k][s]]

There are 6 places (some overlapping) at which s[_][_][_] or k[_][_] matches some subpart of this expression:

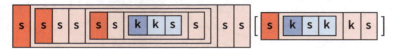

One can see the same thing in the tree form of the expression (the matches are indicated at the roots of their subtrees):

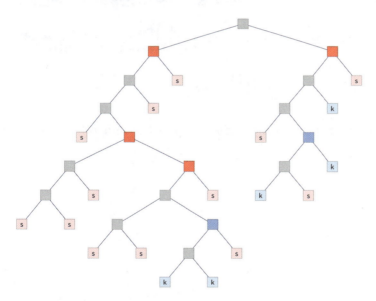

But now the question is: if one's applying combinator rules, which of these matches should one use?

What we've done so far is to follow a particular strategy—usually called leftmost outermost—which can be thought of as looking at the combinator expression as we normally write it out with brackets etc. and applying the first match we encounter in a left-to-right scan, or in this case:

In the Wolfram Language we can find the positions of the matches just using:

In[∘]:= **expr = s[s[s][s][s[s][k[k][s]][s]]][s][s][s[k[s][k]][k][s]]**

In[∘]:= **pos = Position[expr, s[_][_][_] | k[_][_]]**

Out[∘]= **{{0, 0, 0, 1, 1, 0, 1}, {0, 0, 0, 1, 1}, {0, 0, 0, 1}, {0}, {1, 0, 0, 1}, {1}}**

This shows—as above—where these matches are in the expression:

In[∘]:= **expr = s[s[s][s][s[s][k[k][s]][s]]][s][s][s[k[s][k]][k][s]];**

In[∘]:= **pos = Position[expr, s[_][_][_] | k[_][_]];**

In[∘]:= **MapAt[Framed, expr, pos]**

Out[∘]=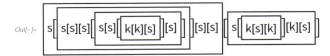

Here are the matches, in the order provided by Position:

{0, 0, 0, 1, 1, 0, 1}	s[s[s][s][s[s][k[k][s]][s]]][s][s][s[k[s][k]][k][s]]
{0, 0, 0, 1, 1}	s[s[s][s][s[s][k[k][s]][s]]][s][s][s[k[s][k]][k][s]]
{0, 0, 0, 1}	s[s[s][s][s[s][k[k][s]][s]]][s][s][s[k[s][k]][k][s]]
{0}	s[s[s][s][s[s][k[k][s]][s]]][s][s] [s[k[s][k]][k][s]]
{1, 0, 0, 1}	s[s[s][s][s[s][k[k][s]][s]]][s][s][s[k[s][k]][k][s]]
{1}	s[s[s][s][s[s][k[k][s]][s]]][s][s][s[k[s][k]][k][s]]

The leftmost-outermost match here is the one with position {0}.

In general the series of indices that specify the position of a subexpression say whether to reach the subexpression one should go left or right at each level as one descends the expression tree. An index 0 says to go to "head", i.e. the f in f[x], or the f[a][b] in f[a][b][c]; an index 1 says to the "first argument", i.e. the x in f[x], or the c in f[a][b][c]. The length of the list of indices gives the depth of the corresponding subexpression.

We'll talk in the next section about how leftmost outermost—and other schemes— are defined in terms of indices. But here the thing to notice is that in our example here Position doesn't give us part {0} first; instead it gives us {0,0,0,1,1,0,1}:

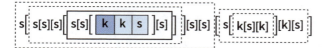

And what's happening is that Position is doing a depth-first traversal of the expression tree to look for matches, so it first descends all the way down the left-hand tree branches—and since it finds a match there, that's what it returns. In the taxonomy we'll discuss in the next section, this corresponds to a leftmost-innermost scheme, though here we'll refer to it as "depth first".

Now consider the example of s[s][s][k[s][s]]. Here is what it does first with the leftmost-outermost strategy we've been using so far, and second with the new strategy:

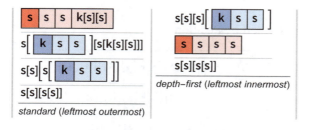

standard (leftmost outermost)

depth–first (leftmost innermost)

There are two important things to notice. First, that in both cases the final result is the same. And second, that the steps taken—and the total number required to get to the final result—is different in the two cases.

Let's consider a larger example: s[s][s][s[s[s]]][k][s]] (**SSS(S(SS))KS**). With our standard strategy we saw above that the evolution of this expression terminates after 89 steps, giving an expression of size 65. With the depth-first strategy the evolution still terminates with the same expression of size 65, but now it takes only 29 steps:

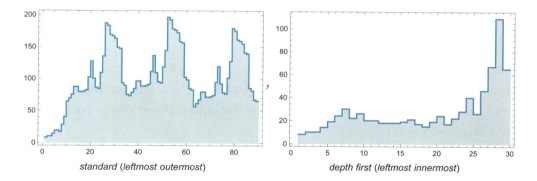

standard (*leftmost outermost*) , depth first (*leftmost innermost*)

It's an important feature of combinator expression evolution that when it terminates—whatever strategy one's used—the result must always be the same. (This
"confluence" property—that we'll discuss more later—is closely related to the
concept of causal invariance in our models of physics.)

What happens when the evolution doesn't terminate? Let's consider the simplest
nonterminating case we found above: s[s][s][s[s]][s][s] (**SSS(SS)SS**). Here's how the
sizes increase with the two strategies we've discussed:

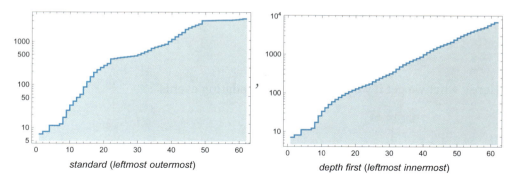

standard (*leftmost outermost*) , depth first (*leftmost innermost*)

The difference is more obvious if we plot the ratios of sizes on successive steps:

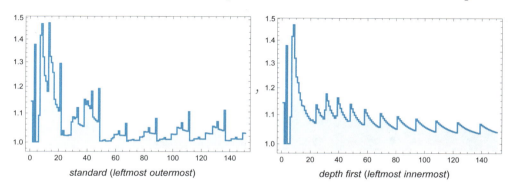

standard (*leftmost outermost*) , depth first (*leftmost innermost*)

In both these pairs of pictures, we can see that the two strategies start off producing the same results, but soon diverge.

OK, so we've looked at two particular strategies for picking which updates to do. But is there a general way to explore all possibilities? It turns out that there is—and it's to use multiway systems, of exactly the kind that are also important in our Physics Project.

The idea is to make a multiway graph in which there's an edge to represent each possible update that can be performed from each possible "state" (i.e. combinator expression). Here's what this looks like for the example of s[s][s][k[s][s]] (**SSS**(**KSS**)) above:

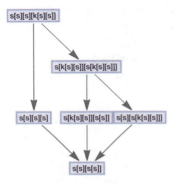

Here's what we get if we include all the "updating events":

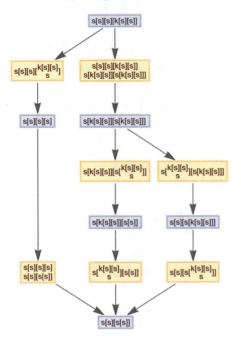

Now each possible sequence of updating events corresponds to a path in the multiway graph. The two particular strategies we used above correspond to these paths:

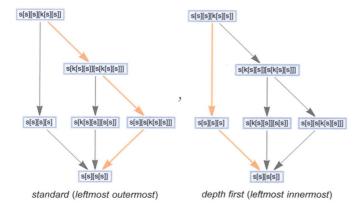

standard (*leftmost outermost*)   depth first (*leftmost innermost*)

We see that even at the first step here, there are two possible ways to go. But in addition to branching, there is also merging, and indeed whichever branch one takes, it's inevitable that one will end up at the same final state—in effect the unique "result" of applying the combinator rules.

Here's a slightly more complicated case, where there starts out being a unique path, but then after 4 steps, there's a branch, but after a few more steps, everything converges again to a unique final result:

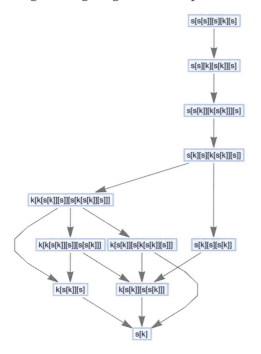

For combinator expressions of size 4, there's never any branching in the multi-way graph. At size 5 the multiway graphs that occur are:

At size 6 the 2688 possible combinator expressions yield the following multiway graphs, with the one shown above being basically as complicated as it gets:

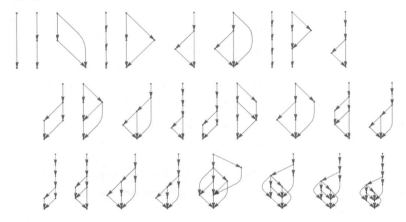

At size 7, much more starts being able to happen. There are rather regular structures like:

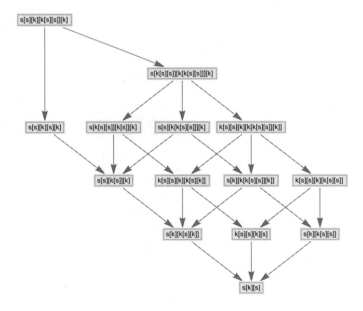

As well as cases like:

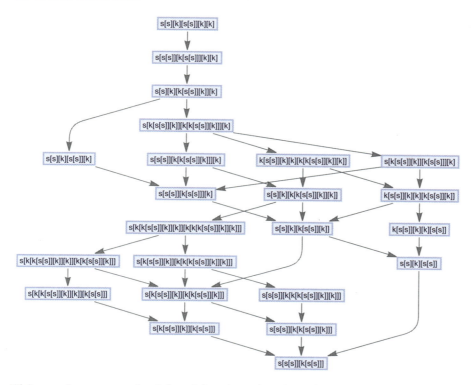

This can be summarized by giving just the size of each intermediate expression, here showing the path defined by our standard leftmost-outermost updating strategy:

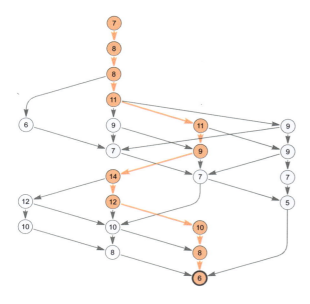

By comparison, here is the path defined by the depth-first strategy above:

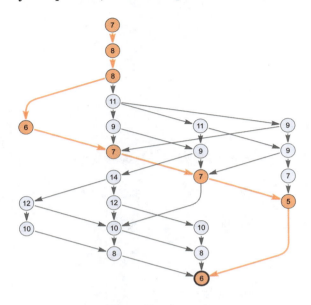

s[s][s][s[s[k]]][k] (**SSS(S(SK))K**) is a case where leftmost-outermost evaluation avoids longer paths and larger intermediate expressions

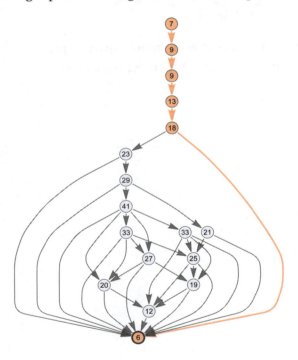

while depth-first evaluation takes more steps:

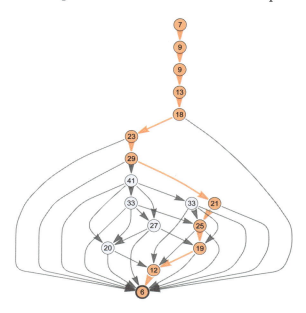

s[s[s]][s][s[s]][s] (**S(SS)S(SS)S**) gives a larger but more uniform multiway graph (s[s[s[s]]][s][s][s] evolves directly to s[s[s]][s][s[s]][s]):

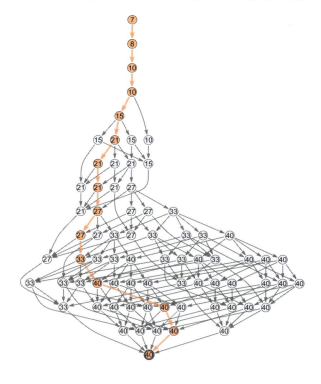

Depth-first evaluation gives a slightly shorter path:

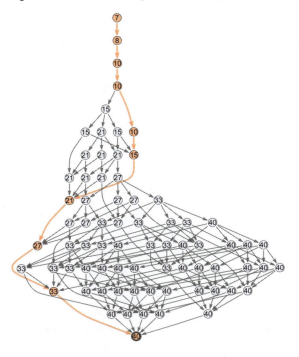

Among size-7 expressions, the largest finite multiway graph (with 94 nodes) is for s[s[s[s]]][s][s][k] (**S(S(SS))SSK**):

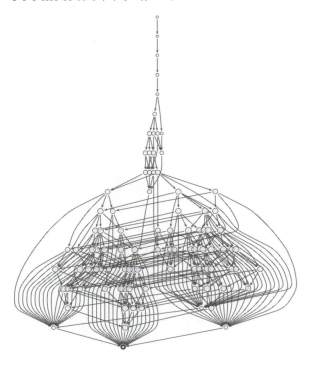

Depending on the path, this can take between 10 and 18 steps to reach its final state:

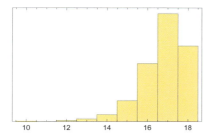

Our standard leftmost-outermost strategy takes 12 steps; the depth first takes 13 steps:

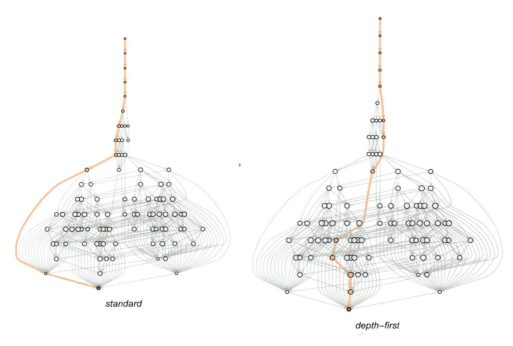

standard   depth-first

But among size-7 combinator expressions there are basically two that do not lead to finite multiway systems: s[s[s]][s][s][s][k] (**S(SS)SSSK**) (which evolves immediately to s[s][s][s[s]][s][k]) and s[s[s]][s][s][s][s] (**S(SS)SSSS**) (which evolves immediately to s[s][s][s[s]][s][s]).

Let's consider s[s[s]][s][s][s][k]. For 8 steps there's a unique path of evolution. But at step 9, the evolution branches

as a result of there being two distinct possible updating events:

Continuing for 14 steps we get a fairly complex multiway system:

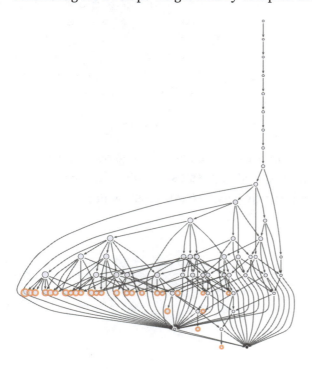

But this isn't "finished"; the nodes circled in red correspond to expressions that are not fixed points, and will evolve further. So what happens with particular evaluation orders?

Here are the results for our two updating schemes:

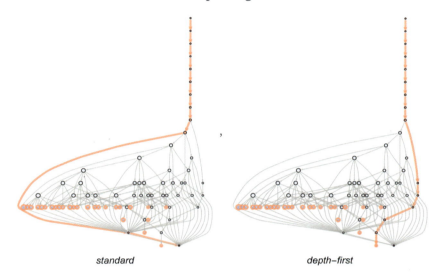

standard   depth–first

Something important is visible here: the leftmost-outermost path leads (in 12 steps) to a fixed-point node, while the depth-first path goes to a node that will evolve further. In other words, at least as far as we can see in this multiway graph, leftmost-outermost evaluation terminates while depth first does not.

There is just a single fixed point visible (s[k]), but there are many "unfinished paths". What will happen with these? Let's look at depth-first evaluation. Even though it hasn't terminated after 14 steps, it does so after 29 steps—yielding the same final result s[k]:

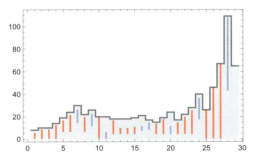

And indeed it turns out to be a general result (known since the 1940s) that if a combinator evolution path is going to terminate, it must terminate in a unique fixed point, but it's also possible that the path won't terminate at all.

Here's what happens after 17 steps. We see more and more paths leading to the fixed point, but we also see an increasing number of "unfinished paths" being generated:

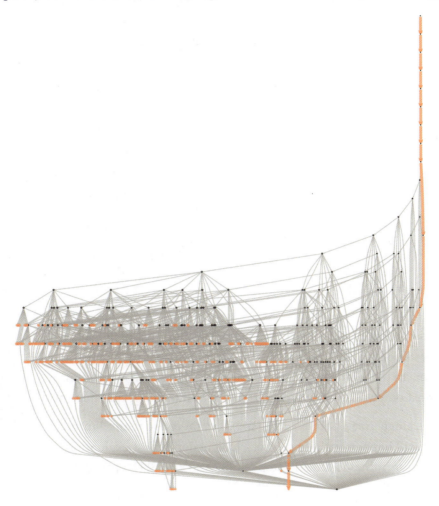

Let's now come back to the other case we mentioned above: s[s[s]][s][s][s][s] (**S(SS)SSSS**). For 12 steps the evolution is unique:

```
s[s[s]][s][s][s][s]
s[s]][s][s[s]][s][s]
s[s[s]][s[s[s]]][s][s]
s[s]][s][s[s]][s]][s]
s[s[s[s]]][s]][s[s[s]]][s]]][s]
s[s]][s][s][s[s[s]]][s]][s]]
s[s]][s][s[s]][s[s[s]]][s]][s]]
s[s[s]][s[s[s]]]][s[s[s]]][s]][s]]
s[s]][s[s[s]]][s]][s[s[s]]][s[s[s]]][s]][s]]]
s[s[s[s]]][s[s[s[s]]][s]][s]][s]][s[s[s]]][s]][s[s[s]]][s[s[s]]][s]][s]]]]
s[s[s[s]]][s[s[s[s]]][s]][s]][s]]][s[s[s]]][s]][s[s[s]]][s[s[s]]][s]][s]]]]]
s[s[s[s]]][s[s[s[s]]][s]][s]][s]]][s[s[s]]][s[s[s[s]]][s[s[s[s]]][s]][s]]][s]]]]][s[s[s[s]]][s[s[s[s]]][s]][s]]]]]]
s[s[s[s]]][s[s[s[s]]][s]][s]][s]]][s[s[s]]][s[s[s[s]]][s[s[s[s]]][s]][s]]][s]]]]][s[s[s[s]]][s[s[s[s]]][s]][s]]]]]][s[s[s[s]]][s[s[s[s]]][s]][s]]]]]]
s[s[s[s]]][s[s[s[s]]][s]][s]][[

   s[s[s[s]]][s[s[s[s]]][s]][s]][s]]]]][s[s[s]]][s[s[s[s]]][s[s[s[s]]][s]][s]]][s]]]][s[s[s[s]]][s[s[s[s]]][s]][s]]]]]][s[s[s[s]]][s[s[s[s]]][s]][s]]]]]]
```

But at that step there are two possible updating events:

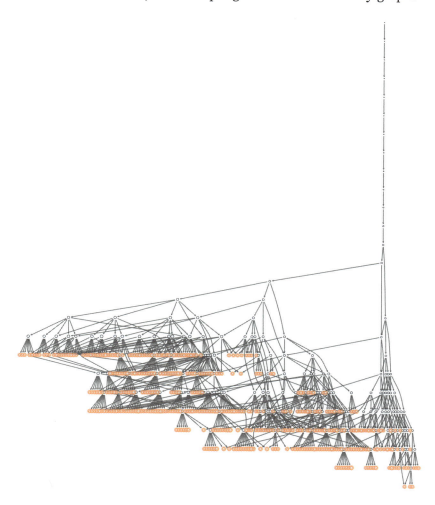

And from there on out, there's rapid growth in the multiway graph:

And what's important here is that there are no fixed points: there is no possible evaluation strategy that leads to a fixed point. And what we're seeing here is an example of a general result: if there is a fixed point in a combinator evolution, then leftmost-outermost evaluation will always find it.

In a sense, leftmost-outermost evaluation is the "most conservative" evaluation strategy, with the least propensity for ending up with "runaway evolution". Its "conservatism" is on display if one compares growth from it and from depth-first evaluation in this case:

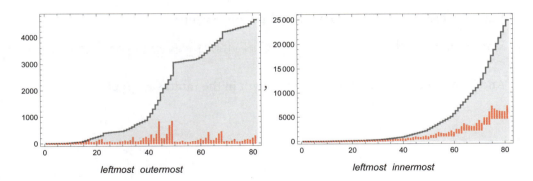

leftmost outermost , *leftmost innermost*

Looking at the multiway graph—as well as others—a notable feature is the presence of "long necks": for many steps every evaluation strategy leads to the same sequence of expressions, and there is just one possible match at each step.

But how long can this go on? For size 8 and below it's always limited (the longest "neck" at size 7 is for s[s[s]][s][s][s][s] and is of length 13; for size 8 it is no longer, but is of length 13 for s[s[s[s]]][s][s][s][s] and k[s[s[s]]][s][s][s][s]]). But at size 9 there are four cases (3 distinct) for which growth continues forever, but is always unique:

{s[s[s[s]]][s[s[s]]]][s], s[s[s[s]]]][s[s[s]]]][s[s]], s[s[s]]][s][s[s[s]][s]][s]], s[s[s]]][k[s[s[s]][s]][s]]}

And as one might expect, all these show rather regular patterns of growth:

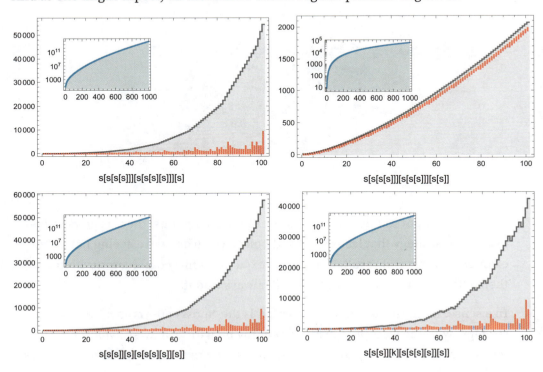

s[s[s[s]]][s[s[s]]]][s] , s[s[s[s]]]][s[s[s]]][s[s]]

s[s[s]]][s][s[s[s]][s]][s]] , s[s[s]]][k[s[s[s]][s]][s]]

The second differences are given in the first and third cases by repeats of (for successive n):

Join[{0, 0, 1}, Table[0, n], {7, 0, 0, 1, 0, 3 $(2^{n+2} - 3)$}]

In the second they are given by repeats of

Join[Table[0, n], {2}]

and in the final case by repeats of:

Join[{0, 1}, Table[0, n], {$-3 \times 2^{n+3} + 18$, $3 \times 2^{n+3} - 11$, 0, 1, 0, $-3 \times 2^{n+3} + 2$, $9 \times 2^{n+2} - 11$}]

The Question of Evaluation Order

As a computational language designer, it's an issue I've been chasing for 40 years: what's the best way to define the order in which one evaluates (i.e. computes) things? The good news is that in a well-designed language (like the Wolfram Language!) it fundamentally doesn't matter, at least much of the time. But in thinking about combinators—and the way they evolve—evaluation order suddenly becomes a central issue. And in fact it's also a central issue in our new model of physics—where it corresponds to the choice of reference frame, for relativity, quantum mechanics and beyond.

Let's talk first about evaluation order as it shows up in the symbolic structure of the Wolfram Language. Imagine you're doing this computation:

In[*]:= **Length[Join[{a, b}, {c, d, e}]]**

Out[*]= 5

The result is unsurprising. But what's actually going on here? Well, first you're computing Join[...]:

In[*]:= **Join[{a, b}, {c, d, e}]**

Out[*]= {a, b, c, d, e}

Then you're taking the result, and providing it as an argument to Length, which then does its job, and gives the result 5. And in general in the Wolfram Language, if you're computing f[g[x]] what'll happen is that x will be evaluated first, followed by g[x], and finally f[g[x]]. (Actually, the head f in f[x] is the very first thing evaluated, and in f[x, y] one evaluates f, then x, then y and then f[x, y].)

And usually this is exactly what one wants, and what people implicitly expect. But there are cases where it isn't. For example, let's say you've defined $x = 1$ (i.e. Set[x, 1]). Now you want to say $x = 2$ (Set[x, 2]). If the x evaluated first, you'd get Set[1, 2], which doesn't make any sense. Instead, you want Set to "hold its first argument", and "consume it" without first evaluating it. And in the Wolfram Language this happens automatically because Set has attribute HoldFirst.

How is this relevant to combinators? Well, basically, the standard evaluation order used by the Wolfram Language is like the depth-first (leftmost-innermost) scheme we described above, while what happens when functions have Hold attributes is like the leftmost-outermost scheme.

But, OK, so if we have something like f[a[x],y] we usually first evaluate a[x], then use the result to compute f[a[x],y]. And that's pretty easy to understand if a[x], say, immediately evaluates to something like 4 that doesn't itself need to be evaluated. But what happens when in f[a[x],y], a[x] evaluates to b[x] which then evaluates to c[x] and so on? Do you do the complete chain of "subevaluations" before you "come back up" to evaluate y, and f[...]?

What's the analog of this for combinators? Basically it's whether when you do an update based on a particular match in a combinator expression, you then just keep on "updating the update", or whether instead you go on and find the next match in the expression before doing anything with the result of the update. The "updating the update" scheme is basically what we've called our depth-first scheme, and it's essentially what the Wolfram Language does in its automatic evaluation process.

Imagine we give the combinator rules as Wolfram Language assignments:

In[∘]:= **s[x_][y_][z_] := x[z][y[z]]**

In[∘]:= **k[x_][y_] := x**

Then—by virtue of the standard evaluation process in the Wolfram Language— every time we enter a combinator expression these rules will automatically be repeatedly applied, until a fixed point is reached:

In[∘]:= **s[s][s][s[s[s]]][k][s]**

Out[∘]= s[s[s[s[s[s[s]]]][k]][k[s[s[s[s[s]]]][k]]]][s[s[s[s[s]]]][k]][k[s[s[s[s[s[s]]]][k]]]][s[s[s[s[s]]]][k]]]][s[s[s[s[s]]]][k][k[s[s[s[s[s]]]][k]]]][s[s[s[s[s]]]][k]]]]][
s[s[s[s[s[s]]]][k]][k[s[s[s[s[s]]]][k]]]][s[s[s[s[s]]]][k]]]]

What exactly is happening "inside" here? If we trace it in a simpler case, we can see that there is repeated evaluation, with a depth-first (AKA leftmost-innermost) scheme for deciding what to evaluate:

In[∘]:= **Dataset[Trace[s[k[k]][k]][s][s]]]**

Out[∘]=

k[k][k]	s[k]	s[k][s]	s[k][s][s]	k[s][s[s]]	s
k					

Of course, given the assignment above for s, if one enters a combinator expression—like s[s][s][s[s]][s][s]—whose evaluation doesn't terminate, there'll be trouble, much as if we define $x = x + 1$ (or $x = \{x\}$) and ask for x. Back when I was first doing language design people often told me that issues like this meant that a language that used automatic infinite evaluation "just couldn't work". But 40+ years later I think I can say with confidence that "programming with infinite evaluation, assuming fixed points" works just great in practice—and in rare cases where there isn't going to be a fixed point one has to do something more careful anyway.

In the Wolfram Language, that's all about specifically applying rules, rather than just having it happen automatically. Let's say we clear our assignments for s and k:

In[ ]:= **Clear[s, k]**

Now no transformations associated with s and k will automatically be made:

In[ ]:= **s[s][s][s[s[s]]][k][s]**

Out[ ]= s[s][s][s[s[s]]][k][s]

But by using /. (ReplaceAll) we can ask that the s, k transformation rules be applied once:

In[ ]:= **s[s][s][s[s[s]]][k][s] /. {s[x_][y_][z_] → x[z][y[z]], k[x_][y_] → x}**

Out[ ]= s[s[s[s]]][s[s[s[s]]]][k][s]

With FixedPointList we can go on applying the rule until we reach a fixed point:

In[ ]:= **FixedPointList[# /. {s[x_][y_][z_] → x[z][y[z]], k[x_][y_] → x} &, s[s][s][s[s[s]]][k][s]]**

```
{s[s][s][s[s[s]]][k][s],
 s[s[s[s]]][s[s[s[s]]]][k][s], s[s[s]][k][s[s[s[s]]][k]][s],
 s[s][s[s[s[s]]][k]][k[s[s[s[s]]][k]]][s],
 s[k[s[s[s[s]]][k]]][s[s[s[s]]][k][k[s[s[s[s]]][k]]]][s],
```

Out[ ]= ⋮

```
 s[s[s[s[s[s[s]]][k]][k[s[s[s[s[s]]]][k]]]]][s[s[s[s[s]]][k]]]],
 s[s[s[s[s[s[s]]]][k]][k[s[s[s[s]]][k]]]]][s[s[s[s]]][k]]][k[s[
     s[s[s[s[s]]][k]][k[s[s[s[s]]][k]]]]][s[s[s[s]]][k]]]]][
     s[s[s[s[s[s]]][k]][k[s[s[s[s]]][k]]]]][s[s[s[s]]][k]]]]}
```

It takes 26 steps—which is different from the 89 steps for our leftmost-outermost evaluation, or the 29 steps for leftmost-innermost (depth-first) evaluation. And, yes, the difference is the result of /. in effect applying rules on the basis of a different scheme than the ones we've considered so far.

But, OK, so how can we parametrize possible schemes? Let's go back to the combinator expression from the beginning of the previous section:

In[ ]:= **s[s[s][s][s[s][k[k][s]][s]]][s][s][s[k[s][k]][k][s]]**

Here are the positions of possible matches in this expression:

In[ ]:= **Position[s[s[s][s][s][s][k[k][s]][s]]][s][s][s[k[s][k]][k][s]], s[_][_][_] | k[_][_]]**

Out[ ]= {{0, 0, 0, 1, 1, 0, 1}, {0, 0, 0, 1, 1}, {0, 0, 0, 1}, {0}, {1, 0, 0, 1}, {1}}

An evaluation scheme must define a way to say which of these matches to actually do at each step. In general we can apply pretty much any algorithm to determine this. But a convenient approach is to think about sorting the list of positions by particular criteria, and then for example using the first k positions in the result.

Given a list of positions, there are two obvious potential types of sorting criteria to use: ones based on the lengths of the position specifications, and ones based on their contents. For example, we might choose (as Sort by default does) to sort shorter position specifications first:

In[ ]:= **Sort[{{0, 0, 0, 1, 1, 0, 1}, {0, 0, 0, 1, 1}, {0, 0, 0, 1}, {0}, {1, 0, 0, 1}, {1}}]**

Out[ ]= {{0}, {1}, {0, 0, 0, 1}, {1, 0, 0, 1}, {0, 0, 0, 1, 1}, {0, 0, 0, 1, 1, 0, 1}}

But what do the shorter position specifications correspond to? They're the more "outer" parts of the combinator expression, higher on the tree. And when we say we're using an "outermost" evaluation scheme, what we mean is that we're considering matches higher on the tree first.

Given two position specifications of the same length, we then need a way to compare these. An obvious one is lexicographic—with 0 sorted before 1. And this corresponds to taking f before x in f[x], or taking the leftmost object first.

We have to decide whether to sort first by length and then by content, or the other way around. But if we enumerate all choices, here's what we get:

leftmost outermost	0	0001	00011	0001101	1	1001
leftmost innermost	0001101	00011	0001	0	1001	1
rightmost outermost	1	1001	0	0001	00011	0001101
rightmost innermost	1001	1	0001101	00011	0001	0
outermost leftmost	0	1	0001	1001	00011	0001101
outermost rightmost	1	0	1001	0001	00011	0001101
innermost leftmost	0001101	00011	0001	1001	0	1
innermost rightmost	0001101	00011	1001	0001	1	0

And here's where the first match with each scheme occurs in the expression tree:

leftmost outermost   *leftmost innermost*   *rightmost outermost*   *rightmost innermost*

outermost leftmost   *outermost rightmost*   *innermost leftmost*   *innermost rightmost*

So what happens if we use these schemes in our combinator evolution? Here's the result for the terminating example s[s][s][s[s[s]]][k][s] above, always keeping only the first match with a given sorting criterion, and at each step showing where the matches were applied:

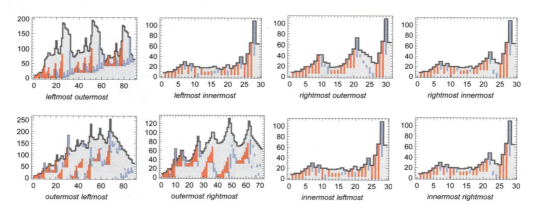

Here now are the results if we allow the first up to 2 matches from each sorted list to be applied:

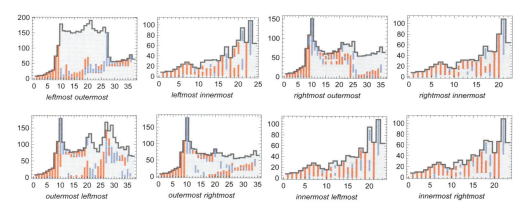

Here are the results for leftmost outermost, allowing up to between 1 and 8 updates at each step:

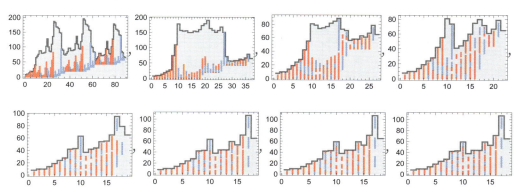

And here's a table of the "time to reach the fixed point" with different evaluation schemes, allowing different numbers of updates at each step:

	1	2	3	4	5	6	7	8
leftmost outermost	89	37	27	22	19	18	18	18
leftmost innermost	29	24	24	21	19	18	18	18
rightmost outermost	31	36	28	21	19	19	18	18
rightmost innermost	29	22	23	23	19	18	18	18
outermost leftmost	89	37	27	22	19	18	18	18
outermost rightmost	71	35	26	20	19	19	18	18
innermost leftmost	29	23	23	23	20	18	18	18
innermost rightmost	29	23	22	25	22	18	18	18

Not too surprisingly, the time to reach the fixed point always decreases when the number of updates that can be done at each step increases.

For the somewhat simpler terminating example s[s[s[s]]][s][s][s] (**S(S(SS))SSS**) we can explicitly look at the updates on the trees at each step for each of the different schemes:

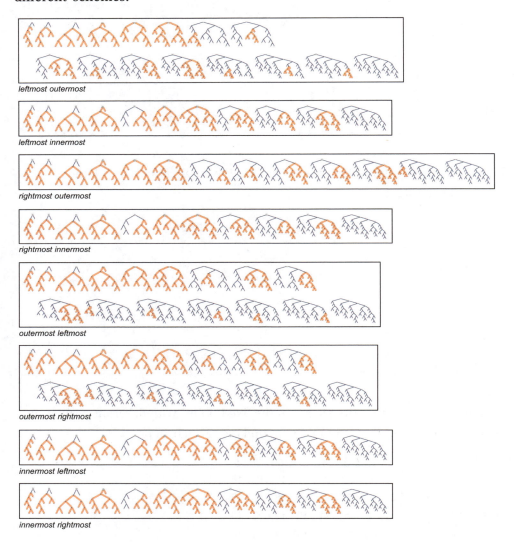

leftmost outermost

leftmost innermost

rightmost outermost

rightmost innermost

outermost leftmost

outermost rightmost

innermost leftmost

innermost rightmost

OK, so what about a combinator expression that does not terminate? What will these different evaluation schemes do? Here are the results for s[s[s]][s][s][s][s] (**S(SS)SSSS**) over the course of 50 steps, in each case using only one match at each step:

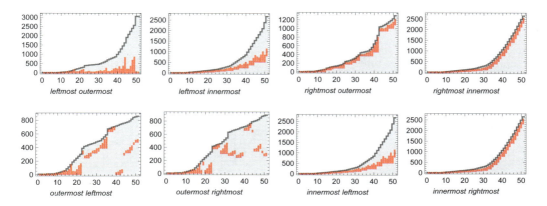

And here is what happens if we allow successively more matches (selected in leftmost-outermost order) to be used at each step:

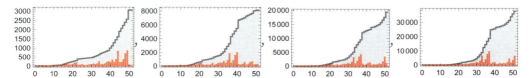

Not surprisingly, the more matches allowed, the faster the growth in size (and, yes, looking at pictures like this suggests studying a kind of "continuum limit" or "mean field theory" for combinator evolution):

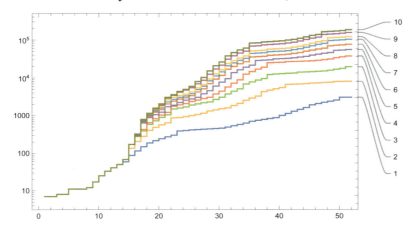

It's interesting to look at the ratios of sizes on successive steps for different updating schemes (still for s[s[s]][s][s][s][s]). Some schemes lead to much more "obviously simple" long-term behavior than others:

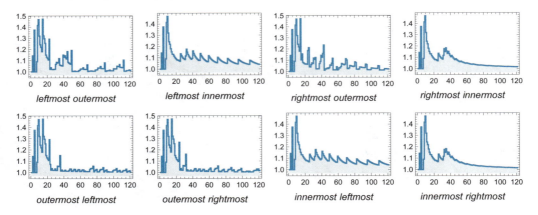

In fact, just changing the number of allowed matches (here for leftmost outermost) can have similar effects:

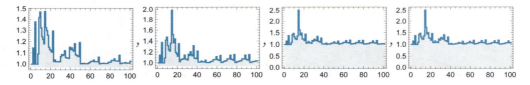

What about for other combinator expressions? Different updating schemes can lead to quite different behavior. Here's s[s[s]][s][s[s[s]]][k] (**S(SS)S(S(SS))K**):

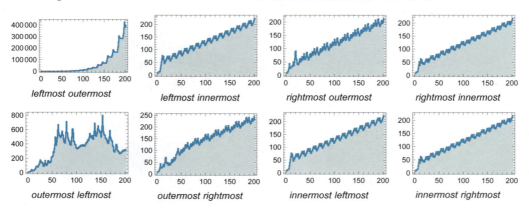

And here's s[s[s]][s][s][s][s[k]] (**S(SS)SSS(SK)**)—which for some updating schemes gives purely periodic behavior (something which can't happen without a k in the original combinator expression):

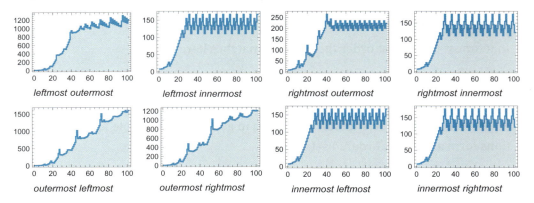

It's worth noting that—at least when there are k's involved—different updating schemes can even change whether the evaluation of a particular combinator expression ever terminates. This doesn't happen below size 8. But at size 8, here's what happens for example with s[s][s][s[s]][s][s][k] (**SSS(SS)SSK**):

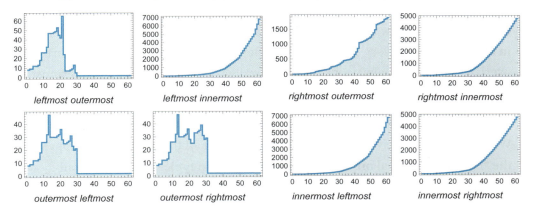

For some updating schemes it reaches a fixed point (always just s[k]) but for others it gives unbounded growth. The innermost schemes are the worst in terms of "missing fixed points"; they do it for 16 size-8 combinator expressions. But (as we mentioned earlier) leftmost outermost has the important feature that it'll never miss a fixed point if one exists—though sometimes at the risk of taking an overly ponderous route to the fixed point.

But so if one's applying combinator-like transformation rules in practice, what's the best scheme to use? The Wolfram Language /. (ReplaceAll) operation in effect uses a leftmost-outermost scheme—but with an important wrinkle: instead of just using one match, it uses as many non-overlapping matches as possible.

Consider again the combinator expression:

s[s[s][s][s[s][k[k][s]][s]]][s][s][s[k[s][k]][k][s]]

In leftmost-outermost order the possible matches here are:

{{0}, {0, 0, 0, 1}, {0, 0, 0, 1, 1}, {0, 0, 0, 1, 1, 0, 1}, {1}, {1, 0, 0, 1}}

But the point is that the match at position {0} overlaps the match at position {0,0,0,1} (i.e. it is a tree ancestor of it). And in general the possible match positions form a partially ordered set, here:

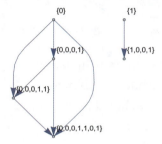

One possibility is always to use matches at the "bottom" of the partial order—or in other words, the very innermost matches. Inevitably these matches can't overlap, so they can always be done in parallel, yielding a "parallel innermost" evaluation scheme that is potentially faster (though runs the risk of not finding a fixed point at all).

What /. does is effectively to use (in leftmost order) all the matches that appear at the "top" of the partial order. And the result is again typically faster overall updating. In the s[s][s][s[s]][s][s][k] example above, repeatedly applying /. (which is what //. does) finds the fixed point in 23 steps, while it takes ordinary one-replacement-at-a-time leftmost-outermost updating 30 steps—and parallel innermost doesn't terminate in this case:

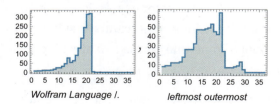

Wolfram Language /.   *leftmost outermost*

For s[s][s][s[s]]][k][s] (**SSS(S(SS))KS**) parallel innermost does terminate, getting a result in 27 steps compared to 26 for /.—but with somewhat smaller intermediate expressions:

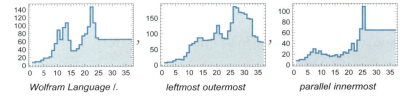

Wolfram Language /.   leftmost outermost   parallel innermost

For a case in which there isn't a fixed point, however, /. will often lead to more rapid growth. For example, with s[s[s]][s][s][s][s] (**S(SS)SSSS**) it basically gives pure exponential $2^{t/2}$ growth (and eventually so does parallel innermost):

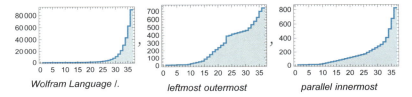

Wolfram Language /.   leftmost outermost   parallel innermost

In *A New Kind of Science* I gave a bunch of results for combinators with /. updating, finding much of the same kind of behavior for "combinators in the wild" as we've seen here.

But, OK, so we've got the updating scheme of /. (and its repeated version //.), and we've got the updating scheme for automatic evaluation (with and without functions with "hold" attributes). But are there other updating schemes that might also be useful, and if so, how might we parametrize them?

I've wondered about this since I was first designing SMP—the forerunner to Mathematica and the Wolfram Language—more than 40 years ago. One place where the issue comes up is in automatic evaluation of recursively defined functions. Say one has a factorial function defined by:

f[1] = 1; f[n_] := n f[n − 1]

What will happen if one asks for f[0]? With the most obvious depth-first evaluation scheme, one will evaluate f[-1], f[-2], etc. forever, never noticing that everything is eventually going to be multiplied by 0, and so the result will be 0. If instead of automatic evaluation one was using //. all would be well—because it's using a different evaluation order:

In[ ]:= **f[0] //. f[n_] → n f[n − 1]**

Out[ ]= 0

Let's consider instead the recursive definition of Fibonacci numbers (to make this more obviously "combinator like" we could for example use Construct instead of Plus):

f[1] = f[2] = 1; f[n_] := f[n−1]+f[n−2]

If you ask for **f[7]** you're essentially going to be evaluating this tree:

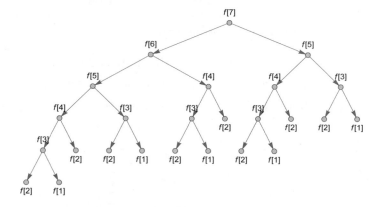

But the question is: how do you do it? The most obvious approach amounts to doing a depth-first scan of the tree—and doing about ϕ^n computations. But if you were to repeatedly use /. instead, you'd be doing more of a breadth-first scan, and it'd take more like $O(n^2)$ computations:

In[ ]:= **FixedPointList[# /. {f[1] → 1, f[2] → 1, f[n_] → f[n−1]+f[n−2]} &, f[7]]**

Out[ ]= {f[7], f[5] + f[6], f[3] + 2 f[4] + f[5], f[1] + f[2] + f[3] + 2 (f[2] + f[3]) + f[4],
          2 + f[1] + 2 f[2] + 2 (1 + f[1] + f[2]) + f[3], 11 + f[1] + f[2], 13, 13}

But how can one parametrize these different kinds of behavior? From our modern perspective in the Wolfram Physics Project, it's like picking different foliations—or different reference frames—in what amount to causal graphs that describe the dependence of one result on others. In relativity, there are some standard reference frames—like inertial frames parametrized by velocity. But in general it's not easy to "describe reasonable reference frames", and we're typically reduced to just talking about named metrics (Schwarzschild, Kerr, ...), much like here we're talking about "named updated orders" ("leftmost inner-most", "outermost rightmost", ...).

But back in 1980 I did have an idea for at least a partial parametrization of evaluation orders. Here it is from section 3.1 of the SMP documentation:

> The simplification of expressions proceeds as follows:
>
> Numbers   Ordinary numbers remain unchanged.
>
> Symbols   A symbol is replaced by the simplified form of any value assigned [3.2] to it.
>
> Projections
>
> 1.   Each filter is simplified in turn, unless the value (k) of any corresponding **Smp** property carried by the projector is **0**. (Future parallel-processing implementations may not respect this ordering.) The simplification of a filter is carried out until its value no longer changes, or until any projectors not carrying property **Rec** have appeared recursively at most k times (see below). If any filter is found to be extended [4] with respect to the projector, then the projection is replaced or encased as specified in the relevant property list [4].
>
> 2.   Projections with **Flat**, **Comm** or **Reor** properties [4, 7.7] are cast into canonical form.

What I called a "projection" then is what we'd call a function now; a "filter" is what we'd now call an argument. But basically what this is saying is that usually the arguments of a function are evaluated (or "simplified" in SMP parlance) before the function itself is evaluated. (Though note the ahead-of-its-time escape clause about "future parallel-processing implementations" which might evaluate arguments asynchronously.)

But here's the funky part: functions in SMP also had Smp and Rec properties (roughly, modern "attributes") that determined how recursive evaluation would be done. And in a first approximation, the concept was that Smp would choose between innermost and outermost, but then in the innermost case, Rec would say how many levels to go before "going outermost" again.

And, yes, nobody (including me) seems to have really understood how to use these things. Perhaps there's a natural and easy-to-understand way to parametrize evaluation order (beyond the /. vs. automatic evaluation vs. hold attributes mechanism in Wolfram Language), but I've never found it. And it's not encouraging here to see all the complexity associated with different updating schemes for combinators.

By the way, it's worth mentioning that there is always a way to completely specify evaluation order: just do something like procedural programming, where every "statement" is effectively numbered, and there can be explicit Goto's that say what statement to execute next. But in practice this quickly gets extremely fiddly and fragile—and one of the great values of functional programming is that it streamlines things by having "execution order" just implicitly determined by the order in which functions get evaluated (yes, with things like Throw and Catch also available).

And as soon as one's determining "execution order" by function evaluation order, things are immediately much more extensible: without having to specify anything else, there's automatically a definition of what to do, for example, when one gets a piece of input with more complex structure. If one thinks about it, there are lots of complex issues about when to recurse through different parts of an expression versus when to recurse through reevaluation. But the good news is that at least the way the Wolfram Language is designed, things in practice normally "just work" and one doesn't have to think about them.

Combinator evaluation is one exception, where, as we have seen, the details of evaluation order can have important effects. And presumably this dependence is in fact connected to why it's so hard to understand how combinators work. But studying combinator evaluation once again inspires one (or at least me) to try to find convenient parametrizations for evaluation order—perhaps now using ideas and intuition from physics.

The World of the S Combinator

In the definitions of the combinators s and k

$\{s[x\_][y\_][z\_] \rightarrow x[z][y[z]], k[x\_][y\_] \rightarrow x\}$

S is basically the one that "builds things up", while K is the one that "cuts things down". And historically, in creating and proving things with combinators, it was important to have the balance of both S and K. But what we've seen above makes it pretty clear that S alone can already do some pretty complicated things.

So it's interesting to consider the minimal case of combinators formed solely from S. For size n (i.e. LeafCount$[n]$), there are

$$\text{CatalanNumber}[n-1] = \frac{\text{Binomial}[2\,n+1, n+1]}{2\,n+1}$$

($\sim \dfrac{4^n}{n^{3/2}}$ for large n) possible such combinators, each of which can be characterized simply in terms of the sequence of bracket openings and closings it involves.

Some of these combinators terminate in a limited time, but above size 7 there are ones that do not:

size	total	nonterminating	fraction
2	1	0	0%
3	2	0	0%
4	5	0	0%
5	14	0	0%
6	42	0	0%
7	132	2	1.5%
8	429	41	9.6%
9	1430	276	19%
10	4862	1481	30%
11	16796	6829	41%
12	58786	29288	50%
13	742900	119946	16%
14	2674440	477885	18%
15	9694845	1870502	19%
16	35357670	7238607	20%

And already there's something weird: the fraction of nonterminating combinator expressions steadily increases with size, then precipitously drops, then starts climbing again:

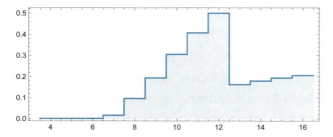

But let's look first at the combinator expressions whose evaluation does terminate. And, by the way, when we're dealing with S alone, there's no possibility of some evaluation schemes terminating and others not: they either all terminate, or none do. (This result was established in the 1930s from the fact that the S combinator—unlike K—in effect "conserves variables", making it an example of the so-called λI calculus.)

With leftmost-outermost evaluation, here are the halting time distributions, showing roughly exponential falloff with gradual broadening:

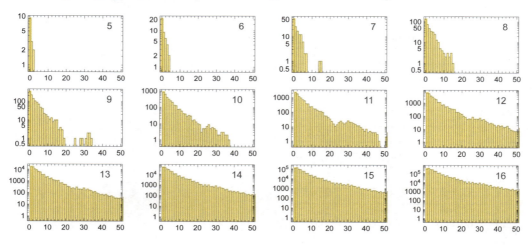

And here are the (leftmost-outermost) "champions"—the combinator expressions that survive longest (with leftmost-outermost evaluation) before terminating:

size	max steps	expression	
2	0	s[s]	**ss**
3	0	s[s][s]	**sss**
4	1	s[s][s][s]	**ssss**
5	2	s[s][s][s][s]	**sssss**
6	4	s[s][s][s][s][s]	**ssssss**
7	15	s[s[s[s]]][s][s][s]	**S(S(SS))SSS**
8	15	s[s[s[s]]][s][s][s]]	**S(S(S(SS))SSS)**
9	86	s[s[s]][s[s]][s[s]][s][s]	**S(SS)(SS)(SS)SS**
10	1109	s[s[s]][s[s]][s[s]][s][s][s]	**S(SSS)(SS)SSSS**
11	1109	s[s[s[s]][s]][s[s]][s[s]][s][s]]	**S(S(SSS)(SS)SSSS)**
12	1444	s[s[s]][s[s]][s[s]][s][s][s][s][s]][s]	**S(SS)(SS)(SSSSSS)S**
13	6317	s[s[s]][s[s]][s[s]][s][s][s][s][s][s]][s]	**S(SS)(SS)(SSSSSSS)S**
14	23 679	s[s[s]][s[s]][s[s]][s][s][s][s][s][s][s]][s]	**S(SS)(SS)(SSSSSSSS)S**
15	131 245	s[s[s]][s[s]][s[s]][s][s][s][s][s][s][s][s]][s]	**S(SS)(SS)(SSSSSSSSS)S**
16	454 708	s[s[s]][s[s]][s[s]][s][s][s][s][s][s][s][s][s]][s]	**S(SS)(SS)(SSSSSSSSSS)S**

The survival (AKA halting) times grow roughly exponentially with size—and notably much slower than what we saw in the SK case above:

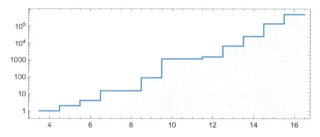

How do the champions actually behave? Here's what happens for a sequence of sizes:

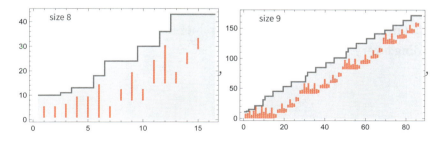

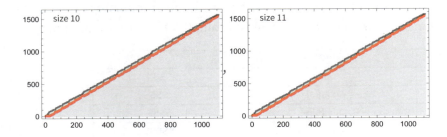

There's progressive increase in size, and then splat: the evolution terminates. Looking at the detailed behavior (here for size 9 with a "right-associative rendering") shows that what's going on is quite systematic:

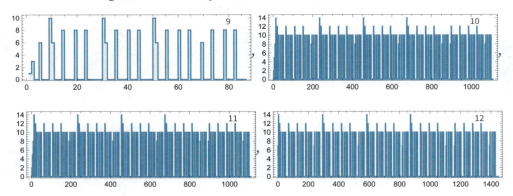

The differences again reflect the systematic character of the behavior:

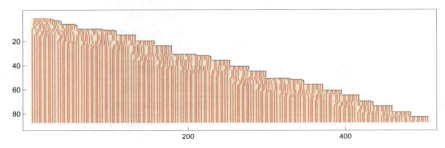

And it seems that what's basically happening is that the combinator is acting as a kind of digital counter that's going through an exponential number of steps—and ultimately building a very regular tree structure:

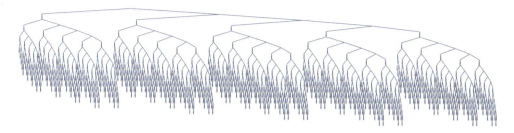

By the way, even though the final state is the same, the evolution is quite different with different evaluation schemes. And for example our "leftmost-outermost champions" actually terminate much faster with depth-first evaluation:

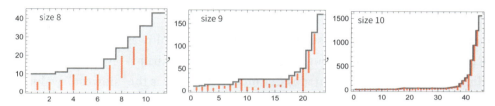

Needless to say, there can be different depth-first (AKA leftmost-innermost) champions, although—somewhat surprisingly—some turn out to be the same (but not sizes 8, 12, 13):

size	max steps	expression	
2	0	s[s]	SS
3	0	s[s][s]	SSS
4	1	s[s][s][s]	SSSS
5	2	s[s][s][s][s]	SSSSS
6	4	s[s][s][s][s][s]	SSSSSS
7	10	s[s[s]]][s][s][s]	S(S(SS))SSS
8	11	s[s[s][s]][s][s][s]	S(SSS)SSSS
9	22	s[s[s]]][s[s]][s[s]][s][s]	S(SS)(SS)(SS)SS
10	44	s[s[s][s]][s[s]][s][s][s]	S(SSS)(SS)SSSS
11	44	s[s[s][s][s]][s[s]][s][s][s]]	S(S(SSS)(SS)SSSS)
12	48	s[s][s[s[s][s]][s[s]]][s][s][s][s]	SS(S(SSS)(SS))SSSS
13	55	s[s[s][s]][s[s[s][s][s][s][s]]][s][s][s]	S(SSS)(SSSSS)SSS

We can get a sense of what happens with all possible evaluation schemes if we look at the multiway graph. Here is the result for the size-8-leftmost-outermost champion s[s[s[s]]][s][s][s]:

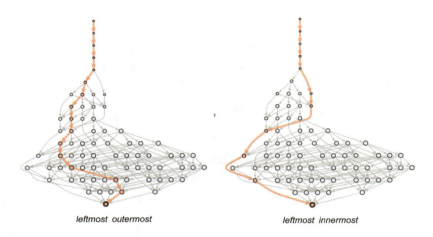

leftmost outermost   *leftmost innermost*

The number of expressions at successive levels in the multiway graph starts off growing quite exponentially, but after 12 steps it rapidly drops—eventually yielding a finite graph with 74 nodes (leftmost outermost is the "slowest" evaluation scheme—taking the maximum 15 steps possible):

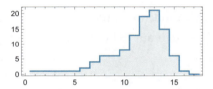

Even for the size-9 champion the full multiway graph is too large to construct explicitly. After 15 steps the number of nodes has reached 6598, and seems to be increasingly roughly like 2^t—even though after at most 86 steps all "dangling ends" must have resolved, and the system must reach its fixed point:

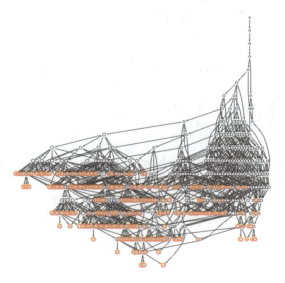

What happens with s combinator expressions that do not terminate? We already saw above some examples of the kind of growth in size one observes (say with leftmost-outermost evaluation). Here are examples with roughly exponential behavior, with differences between successive steps shown on a log scale:

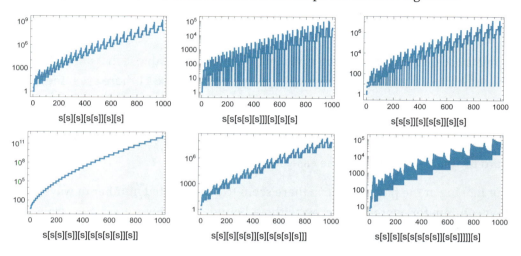

And here are examples of differences shown on a linear scale:

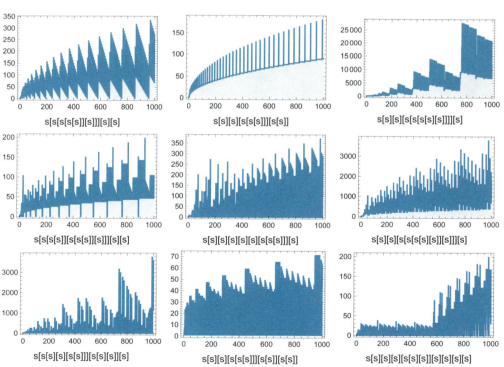

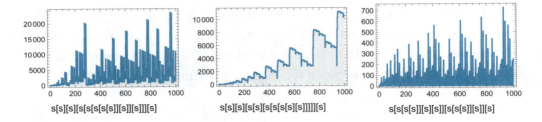

s[s][s][s][s[s[s[s]][s]][s]]][s]   s[s][s][s][s][s[s[s[s]]]]]][s]   s[s[s[s]]][s][s]][s[s[s]][s]][s]

Sometimes there are fairly long transients, but what's notable is that among all the 8629 infinite-growth combinator expressions up to size 11 there are none whose evolution seems to show long-term irregularity in overall size. Of course, something like rule 30 also doesn't show irregularity in overall size; one has to look "inside" to see complex behavior—and difficulties of visualization make that hard to systematically do in the case of combinators.

But looking at the pictures above there seem to be a "limited number of ways" that combinator expressions grow without bound. Sometimes it's rather straight-forward to see how the infinite growth happens. Here's a particularly "pure play" example: the size-9 case s[s[s[s]]][s[s[s]]][s[s]] (**S(S(SS))(S(SS))(SS)**) which evolves the same way with all evaluation schemes (in the pictures, the root of the match at each step is highlighted):

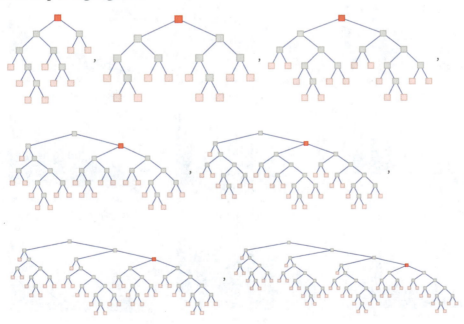

Looking at the subtree "below" each match we see

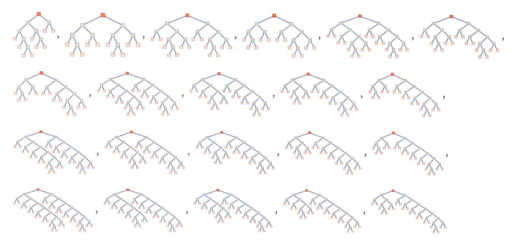

and it is clear that there is a definite progression which will keep going forever, leading to infinite growth.

But if one looks at the corresponding sequence of subtrees for a case like the smallest infinite-growth combinator expression s[s][s][s[s]][s][s] (**SSS(SS)SS**), it's less immediately obvious what's going on:

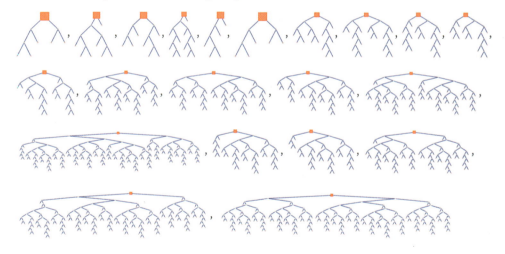

But there's a rather remarkable result from the end of the 1990s that gives one a way to "evaluate" combinator expressions, and tell whether they'll lead to infinite growth—and in particular to be able to say directly from an initial combinator expression whether it'll continue evolving forever, or will reach a fixed point.

One starts by writing a combinator expression like s[s[s[s]]][s[s[s]]][s[s]] (**S**(**S**(**SS**))(**S**(**SS**))(**SS**)) in an explicitly "functional" form:

f[f[f[s, f[s, f[s, s]]], f[s, f[s, s]]], f[s, s]]

Then one imagines f[x, y] as being a function with explicit (say, integer) values. One replaces s by some explicit value (say an integer), then defines values for f[1, 1], f[1, 2], etc.

As a first example, let's say that we take s = 1 and f[x_, y_] = x + y. Then we can "evaluate" the combinator expression above as

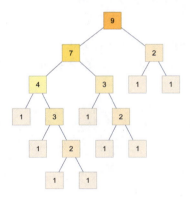

and in this case the value at the root just counts the total size (i.e. LeafCount).

But by changing f one can probe other aspects of the combinator expression tree. And what was discovered in 2000 is that there's a complete way to test for infinite growth by setting up 39 possible values, and making f[x, y] be a particular ("tree automaton") "multiplication table" for these values:

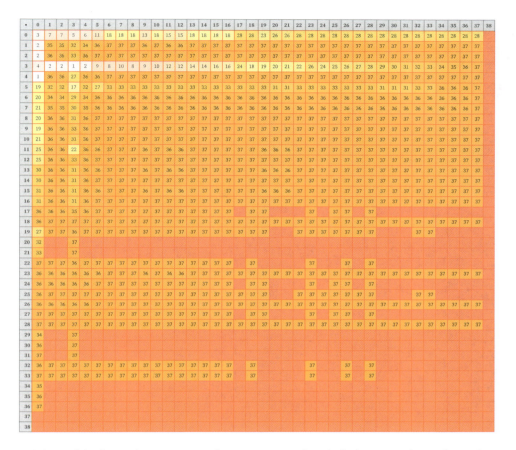

Bright red (value 38) represents the presence of an infinite growth seed—and once one exists, f makes it propagate up to the root of the tree. And with this setup, if we replace s by the value 0, the combinator expression above can be "evaluated" as:

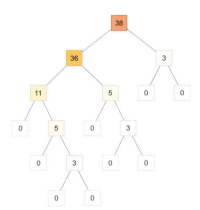

At successive steps in the evolution we get:

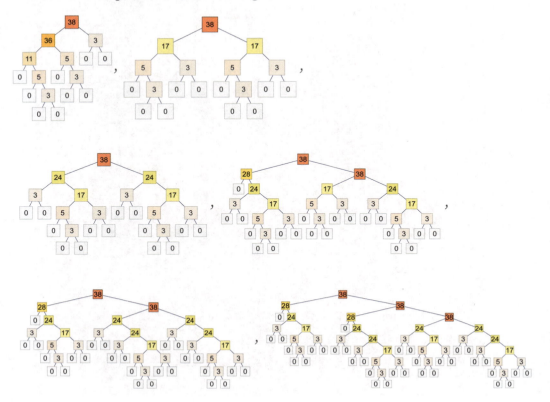

Or after 8 steps:

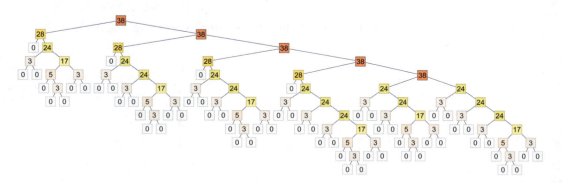

The "lowest 38" is always at the top of the subtree where the match occurs, serving as a "witness" of the fact that this subtree is an infinite growth seed.

Here are some sample size-7 combinator expressions, showing how the two that lead to infinite growth are identified:

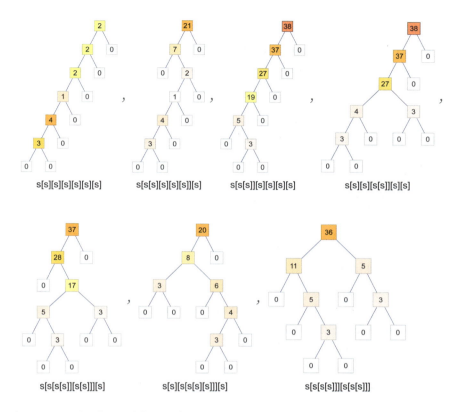

s[s][s][s][s][s][s]   s[s[s][s][s][s]][s]   s[s[s]][s][s][s][s]   s[s][s][s[s]][s][s]

s[s[s[s]][s[s]]][s]   s[s][s[s[s]][s]]][s]   s[s[s[s]]][s[s[s]]]

If we were dealing with combinator expressions involving both S and K we know that it's in general undecidable whether a particular expression will halt. So what does it mean that there's a decidable way to determine whether an expression involving only S halts?

One might assume it's a sign that S alone is somehow computationally trivial. But there's more to this issue. In the past, it has often been thought that a "computation" must involve starting with some initial ("input") state, then ending up at a fixed point corresponding to a final result. But that's certainly not how modern computing in practice works. The computer and its operating system do not completely stop when a particular computation is finished. Instead, the computer keeps running, but the user is given a signal to come and look at something that provides the output for the computation.

There's nothing fundamentally different about how computation universality works in a setup like this; it's just a "deployment" issue. And indeed the simplest possible examples of universality in cellular automata and Turing machines have been proved this way.

So how might this work for S combinator expressions? Basically any sophisticated computation has to live on top of an infinite combinator growth process. Or, put another way, the computation has to exist as some kind of "transient" of potentially unbounded length, that in effect "modulates" the infinite growth "carrier".

One would set up a program by picking an appropriate combinator expression from the infinite collection that lead to infinite growth. Then the evolution of the combinator expression would "run" the program. And one would use some computationally bounded process (perhaps a bounded version of a tree automaton) to identify when the result of the computation is ready—and one would "read it out" by using some computationally bounded "decoder".

My experience in the computational universe—as captured in the Principle of Computational Equivalence—is that once the behavior of a system is not "obviously simple", the system will be capable of sophisticated computation, and in particular will be computation universal. The S combinator is a strange and marginal case. At least in the ways we have looked at it here, its behavior is not "obviously simple". But we have not quite managed to identify things like the kind of seemingly random behavior that occurs in a system like rule 30, that are a hallmark of sophisticated computation, and probably computation universality.

There are really two basic possibilities. Either the S combinator alone is capable of sophisticated computation, and there is, for example, computational irreducibility in determining the outcome of a long S combinator evolution. Or the S combinator is fundamentally computationally reducible—and there is some approach (and maybe some new direction in mathematics) that "cracks it open", and allows one to readily predict everything that an S combinator expression will do.

I'm not sure which way it's going to go—although my almost-uniform experience over the last four decades has been that when I think some system is "too simple" to "do anything interesting" or show sophisticated computation, it eventually proves me wrong, often in bizarre and unexpected ways. (In the case of the S combinator, a possibility—like I found for example in register machines—is that sophisticated computation might first reveal itself in very subtle effects, like seemingly random off-by-one patterns.)

But whatever happens, it's amazing that 100 years after the invention of the S combinator there are still such big mysteries about it. In his original paper, Moses Schönfinkel expressed his surprise that something as simple as S and K were sufficient to achieve what we would now call universal computation. And it will be truly remarkable if in fact one can go even further, and S alone is sufficient: a minimal example of universal computation hiding in plain sight for a hundred years.

(By the way, in addition to ordinary "deterministic" combinator evolution with a particular evaluation scheme, one can also consider the "nondeterministic" case corresponding to all possible paths in the multiway graph. And in that case there's a question of categorizing infinite graphs obtained by nonterminating S combinator expressions—perhaps in terms of transfinite numbers.)

Causal Graphs and the Physicalization of Combinators

Not long ago one wouldn't have had any reason to think that ideas from physics would relate to combinators. But our Wolfram Physics Project has changed that. And in fact it looks as if methods and intuition from our Physics Project—and the connections they make to things like relativity—may give some interesting new insights into combinators, and may in fact make their operation a little less mysterious.

In our Physics Project we imagine that the universe consists of a very large number of abstract elements ("atoms of space") connected by relations—as represented by a hypergraph. The behavior of the universe—and the progression of time—is then associated with repeated rewriting of this hypergraph according to a certain set of (presumably local) rules.

It's certainly not the same as the way combinators work, but there are definite similarities. In combinators, the basic "data structure" is not a hypergraph, but a binary tree. But combinator expressions evolve by repeated rewriting of this tree according to rules that are local on the tree.

There's a kind of intermediate case that we've often used as a toy model for aspects of physics (particularly quantum mechanics): string substitution systems. A combinator expression can be written out "linearly" (say as s[s][s][s[s[s]]][k][s]), but really it's tree-structured and hierarchical. In a string substitution system, however, one just has plain strings, consisting of sequences of characters, without any hierarchy. The system then evolves by repeatedly rewriting the string by applying some local string substitution rule.

For example, one could have a rule like {"A"→"BBB","BB"→"A"}. And just like with combinators, given a particular string—like "BBA"—there are different possible choices about where to apply the rule. And—again like with combinators—we can construct a multiway graph to represent all possible sequences of rewritings:

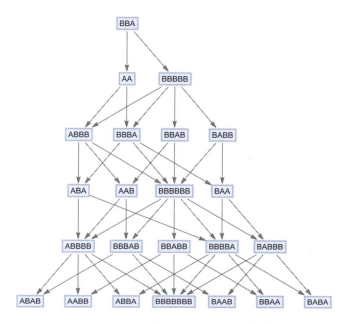

And again as with combinators we can define a particular "evaluation order" that determines which of the possible updates to the string to apply at each step—and that defines a path through the multiway graph.

For strings there aren't really the same notions of "innermost" and "outermost", but there are "leftmost" and "rightmost". Leftmost updating in this case would give the evolution history

{BBA, AA, BBBA, ABA, BBBBA, ABBA, BBBBBA, ABBBA, BBBBBBA, ABBBBA, BBBBBBBA}

which corresponds to the path:

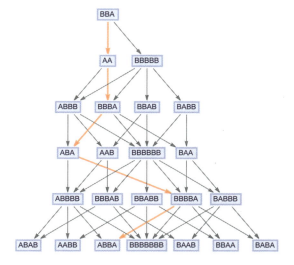

Here's the underlying evolution corresponding to that path, with the updating events indicated in yellow:

But now we can start tracing the "causal dependence" of one event on another. What characters need to have been produced as "output" from a preceding event in order to provide "input" to a new event? Let's look at a case where we have a few more events going on:

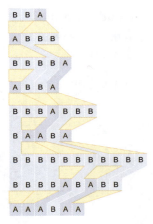

But now we can draw a causal graph that shows causal relationships between events, i.e. which events have to have happened in order to enable subsequent events:

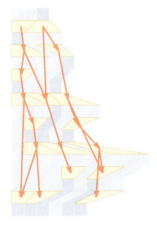

And at a physics level, if we're an observer embedded in the system, operating according to the rules of the system, all we can ultimately "observe" is the "disembodied" causal graph, where the nodes are events, and the edges represent the causal relationships between these events:

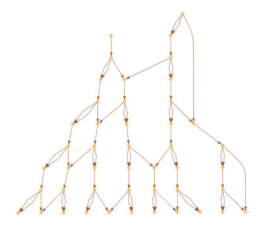

So how does this relate to combinators? Well, we can also create causal graphs for those—to get a different view of "what's going on" during combinator evolution.

There is significant subtlety in exactly how "causal dependence" should be defined for combinator systems (when is a copied subtree "different"?, etc.). Here I'll use a straightforward definition that'll give us an indication of how causal relationships in combinators work, but that's going to require further refinement to fit in with other definitions we want.

Imagine we just write out combinator expressions in a linear way. Then here's a combinator evolution:

To understand causal relationships we need to trace "what gets rewritten to what"—and which previous rewriting events a given rewriting event "takes its input from". It's helpful to look at the rewriting process above in terms of trees:

Going back to a textual representation, we can show the evolution in terms of "states", and the "events" that connect them. Then we can trace (in orange) what the causal relationships between the events are:

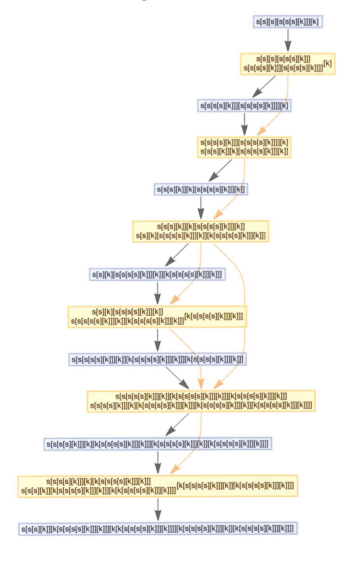

Continuing this for a few more steps we get:

Now keeping only the causal graph, and continuing until the combinator evolution terminates, we get:

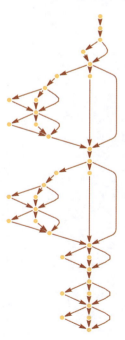

It's interesting to compare this with a plot that summarizes the succession of rewriting events:

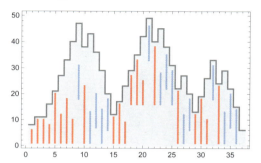

So what are we actually seeing in the causal graph? Basically it's showing us what "threads of evaluation" occur in the system. When there are different parts of the combinator expression that are in effect getting updated independently, we see multiple causal edges running in parallel. But when there's a synchronized evaluation that affects the whole system, we just see a single thread—a single causal edge.

The causal graph is in a sense giving us a summary of the structure of the combinator evolution, with many details stripped out. And even when the size of the combinator expression grows rapidly, the causal graph can still stay quite simple. So, for example, the growing combinator s[s][s][s[s]][s][s] has a causal graph that forms a linear chain with simple "side loops" that get systematically further apart:

Sometimes it seems that the growth dies out because different parts of the combinator system become causally disconnected from each other:

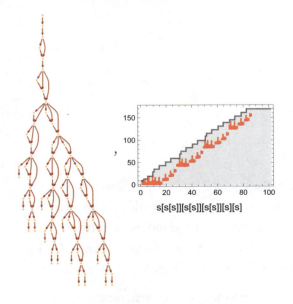

s[s[s]][s[s]][s[s]][s][s]

Here are a few other examples:

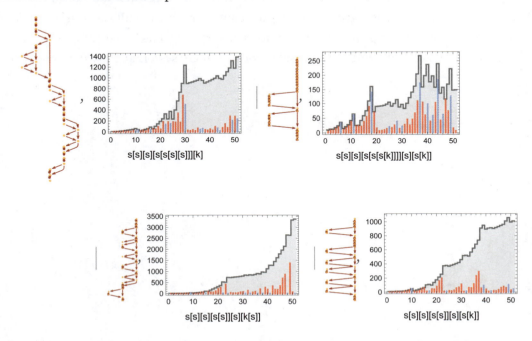

s[s][s][s[s[s]]][k]   s[s][s][s[s[s[k]]]][s][s[k]]

s[s][s][s[s]][s][k[s]]   s[s][s][s[s]][s][s[k]]

But do such causal graphs depend on the evaluation scheme used? This turns out to be a subtle question that depends sensitively on definitions of identity for abstract expressions and their subexpressions.

The first thing to say is that combinators are confluent, in the sense that different evaluation schemes—even if they take different paths—must always give the same final result whenever the evolution of a combinator expression terminates. And closely related to this is the fact that in the multiway graph for a combinator system, any branching must be accompanied by subsequent merging.

For both string and hypergraph rewriting rules, the presence of these properties is associated with another important property that we call causal invariance. And causal invariance is precisely the property that causal graphs produced by different updating orders must always be isomorphic. (And in our model of physics, this is what leads to relativistic invariance, general covariance, objective measurement in quantum mechanics, etc.)

So is the same thing true for combinators? It's complicated. Both string and hypergraph rewriting systems have an important simplifying feature: when you update something in them, it's reasonable to think of the thing you update as being "fully consumed" by the updating event, with a "completely new thing" being created as a result of the event.

But with combinators that's not such a reasonable picture. Because when there's an updating event, say for s[x][y][z], x can be a giant subtree that you end up "just copying", without, in a sense, "consuming" and "reconstituting". In the case of strings and hypergraphs, there's a clear distinction between elements of the system that are "involved in an update", and ones that aren't. But in a combinator system, it's not so obvious whether nodes buried deep in a subtree that's "just copied" should be considered "involved" or not.

There's a complicated interplay with definitions used in constructing multiway graphs. Consider a string rewriting system. Start from a particular state and then apply rewriting rules in all possible ways:

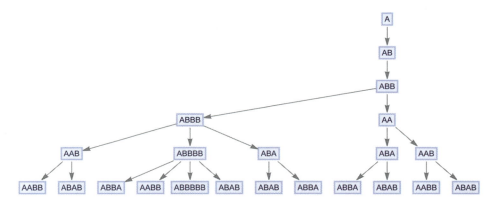

Absent anything else, this will just generate a tree of results. But the crucial idea behind multiway graphs is that when states are identical, they should be merged, in this case giving:

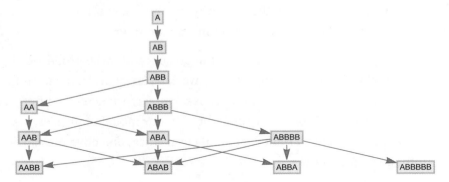

For strings it's very obvious what "being identical" means. For hypergraphs, the natural definition is hypergraph isomorphism. What about for combina-tors? Is it pure tree isomorphism, or should one take into account the "prov-enance" of subtrees?

(There are also questions like whether one should define the nodes in the multi-way graph in terms of "instantaneous states" at all, or whether instead they should be based on "causal graphs so far", as obtained with particular event histories.)

These are subtle issues, but it seems pretty clear that with appropriate definitions combinators will show causal invariance, so that (appropriately defined) causal graphs will be independent of evaluation scheme.

By the way, in addition to constructing causal graphs for particular evolution histories, one can also construct multiway causal graphs representing all possible causal relationships both within and between different branches of history. This shows the multiway graph for the (terminating) evolution of s[s][s][s[s[k]]][k], annotated with casual edges:

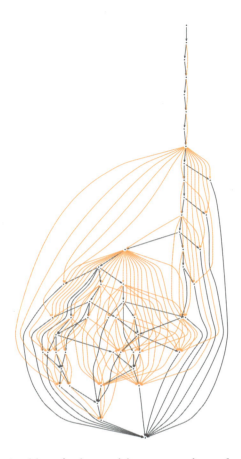

And here's the multiway causal graph alone in this case:

(And, yes, the definitions don't all quite line up here, so the individual instances of causal graphs that can be extracted here aren't all the same, as causal invariance would imply.)

The multiway causal graph for s[s[s]][s][s][s][s] shows a veritable explosion of causal edges:

In our model of physics, the causal graph can be thought of as a representation of the structure of spacetime. Events that follow from each other are "timelike separated". Events that can be arranged so that none are timelike separated can be considered to form a "spacelike slice" (or a "surface of simultaneity"), and to be spacelike separated. (Different foliations of the causal graph correspond to different "reference frames" and identify different sets of events as being in the same spacelike slice.)

When we're dealing with multiway systems it's also possible for events to be associated with different "threads of history"—and so to be branchlike separated. But in combinator systems, there's yet another form of separation between events that's possible—that we can call "treelike separation".

Consider these two pairs of updating events:

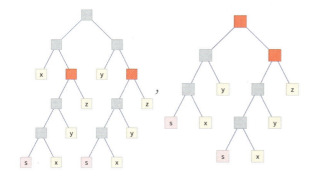

In the first case, the events are effectively "spacelike separated". They are connected by being in the same combinator expression, but they somehow appear at "distinct places". But what about the second case? Again the two events are connected by being in the same combinator expression. But now they're not really "at distinct places"; they're just "at distinct scales" in the tree.

One feature of hypergraph rewriting systems is that in large-scale limits the hypergraphs they produce can behave like continuous manifolds that potentially represent physical space, with hypergraph distances approximating geometric distances. In combinator systems there is almost inevitably a kind of nested structure that may perhaps be reminiscent of scale-invariant critical phenomena and ideas like scale relativity. But I haven't yet seen combinator systems whose limiting behavior produces something like finite-dimensional "manifold-like" space.

It's common to see "event horizons" in combinator causal graphs, in which different parts of the combinator system effectively become causally disconnected. When combinators reach fixed points, it's as if "time is ending"—much as it does in spacelike singularities in spacetime. But there are no doubt new "treelike" limiting phenomena in combinator systems, that may perhaps be reflected in properties of hyperbolic spaces.

One important feature of both string and hypergraph rewriting systems is that their rules are generally assumed to be somehow local, so that the future effect of any given element must lie within a certain "cone of influence". Or, in other words, there's a light cone which defines the maximum spacelike separation of events that can be causally connected when they have a certain timelike separation. In our model of physics, there's also an "entanglement cone" that defines maximum branchlike separation between events.

But what about in combinator systems? The rules aren't really "spatially local", but they are "tree local". And so they have a limited "tree cone" of influence, associated with a "maximum treelike speed"—or, in a sense, a maximum speed of scale change.

Rewriting systems based on strings, hypergraphs and combinator expressions all have different simplifying and complexifying features. The relation between underlying elements ("characters arranged in sequence") is simplest for strings. The notion of what counts as the same element is simplest for hypergraphs. But the relation between the "identities of elements" is probably simplest for combinator expressions.

Recall that we can always represent a combinator expression by a DAG in which we "build up from atoms", sharing common subexpressions all the way up:

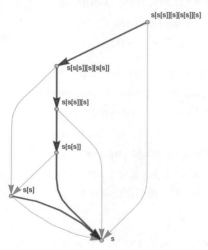

But what does combinator evolution look like in this representation? Let's start from the extremely simple case of k[x][y], which in one step becomes just x. Here's how we can represent this evolution process in DAGs:

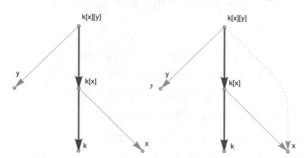

The dotted line in the second DAG indicates an update event, which in this case transforms k[x][y] to the "atom" x.

Now let's consider s[x][y][z]. Once again there's a dotted line that signifies the evolution:

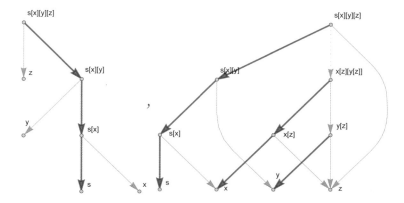

Now let's add an extra wrinkle: consider not k[x][y] but s[k[x][y]]. The outer s doesn't really do anything here. But it still has to be accounted for, in the sense that it has to be "wrapped back around" the x that comes from k[x][y] → x. We can represent that "rewrapping" process, by a "tree pullback pseudoevent" indicated by the dotted line:

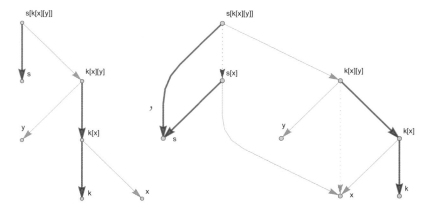

If a given event happens deep inside a tree, there'll be a whole sequence of "pullback pseudoevents" that "reconstitute the tree".

Things get quite complicated pretty quickly. Here's the (leftmost-outermost) evolution of s[s[s]][s][k][s] to its fixed point in terms of DAGs:

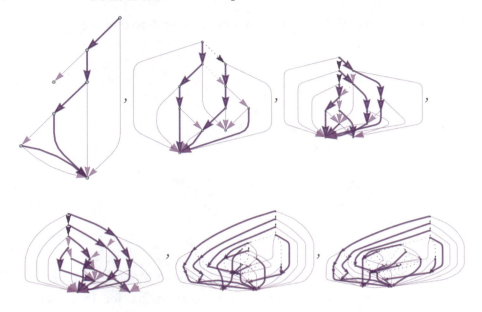

Or with labels:

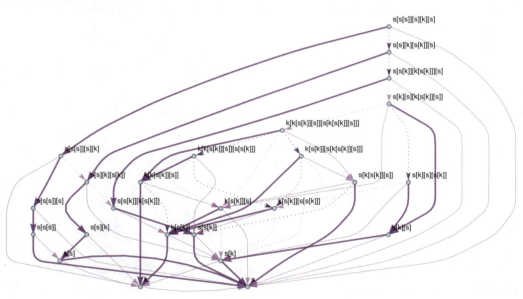

One notable feature is that this final DAG in a sense encodes the complete history of the evolution—in a "maximally shared" way. And from this DAG we can construct a causal graph—whose nodes are derived from the edges in the

DAG representing update events and pseudoevents. It's not clear how to do this in the most consistent way—particularly when it comes to handling pseudo-events. But here's one possible version of a causal graph for the evolution of s[s[s]][s][k][s] to its fixed point—with the yellow nodes representing events, and the gray ones pseudoevents:

Combinator Expressions as Dynamical Systems

Start with all possible combinator expressions of a certain size, say involving only s. Some are immediately fixed points. But some only evolve to fixed points. So how are the possible fixed points distributed in the set of all possible combinator expressions?

For size 6 there are 42 possible combinator expressions, and all evolve to fixed points—but only 27 distinct ones. Here are results for several combinator sizes:

n	1	2	3	4	5	6	7	8	9	10	11	12
all	1	1	2	5	14	42	132	429	1430	4862	16 796	58 786
terminating	1	1	2	5	14	42	130	388	1154	3381	9967	29 498
distinct	1	1	2	4	10	27	77	213	592	1637	4574	12 899
fraction distinct	1	1	1	0.8	0.71	0.64	0.59	0.55	0.51	0.48	0.46	0.44

As the size of the combinator expression goes up, the fraction of distinct fixed points seems to systematically go down:

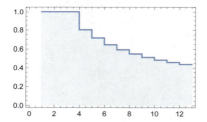

And what this shows is that combinator evolution is in a sense a "contractive" process: starting from all possible expressions, there's only a certain "attractor" of expressions that survives. Here's a "state transition graph" for initial expressions of size 9 computed with leftmost-outermost evaluation (we'll see a more general version in the next section):

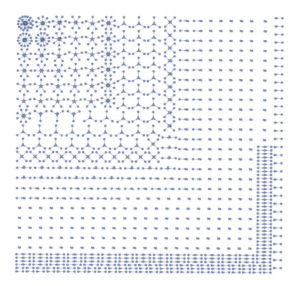

This shows the prevalence of different fixed-point sizes as a function of the size of the initial expression:

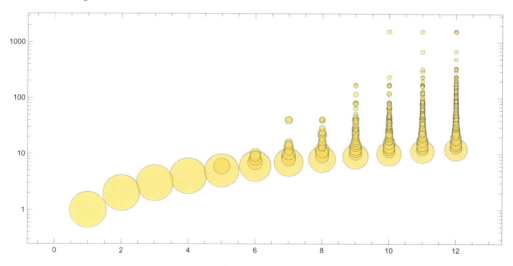

What about the cases that don't reach fixed points? Can we somehow identify different equivalent classes of infinite combinator evolutions (perhaps analogously to the way we can identify different transfinite numbers)? In general we can look at similarities between the multiway systems that are generated, since these are always independent of updating scheme (see the next section).

But something else we can do for both finite and infinite evolutions is to consider the set of subexpressions common to different steps in the evolution—or across different evolutions. Here's a plot of the number of copies of the ultimately most frequent subexpressions at successive steps in the (leftmost-outermost) evolution of s[s][s][s[s]][s][s] (**SSS(SS)SS**):

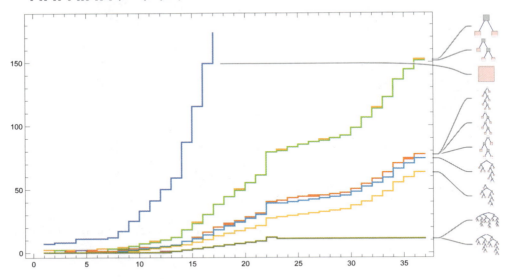

The largest subexpression shown here has size 29. And as the picture makes clear, most subexpressions do not appear with substantial frequency; it's only a thin set that does.

Looking at the evolution of all possible combinator expressions up to size 8, one sees gradual "freezing out" of certain subexpressions (basically as a result of their involvement in halting), and continued growth of others:

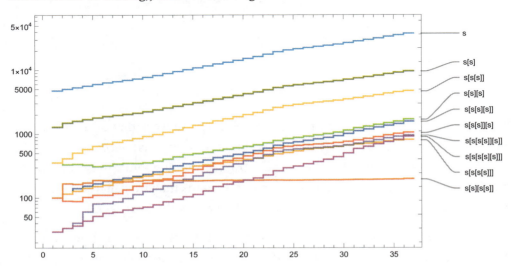

In an attempt to make contact with traditional dynamical systems theory it's interesting to try to map combinator expressions to numbers. A straightforward way to do this (particularly when one's only dealing with expressions involving s) is to use Polish notation, which represents

s[s[s]][s[s[s[s]][s]][s]][s]][s[s[s[s]][s[s[s[s]][s]][s]]]][s[s[s[s]][s[s[s[s]][s]][s]][s]]]

as

••••s•ss••s••s•ssss•s••s•ss••s••s•ssss•s••s•ss••s••s•ssss

or the binary number

11110100110110100001011010011011010000101101001101101000

i.e., in decimal:

137 839 369 892 767 440

Represented in terms of numbers like this, we can plot all subexpressions which arise in the evolution of s[s][s][s[s]][s][s] (**SSS(SS)SS**):

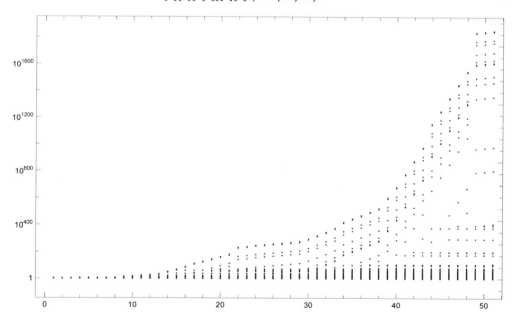

Making a combined picture for all combinator expressions up to size 8, one gets:

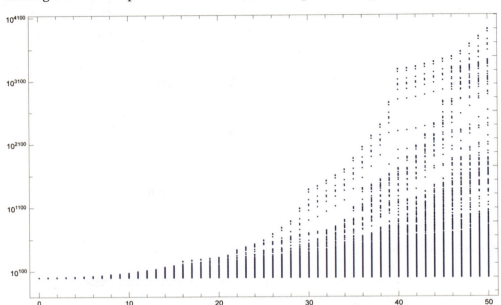

There's definitely some structure: one's not just visiting every possible subexpression. But quite what the limiting form of this might be is not clear.

Another type of question to ask is what the effect of a small change in a combinator expression is on its evolution. The result will inevitably be somewhat subtle—because there is both spacelike and treelike propagation of effects in the evolution.

As one example, though, consider evolving s[s][s][s[s]][s][s] (**SSS(SS)SS**) for 20 steps (to get an expression of size 301). Now look at the effect of changing a single s in this expression to s[s], and then evolving the result. Here are the sizes of the expressions that are generated:

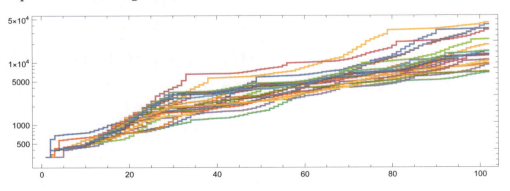

Equality and Theorem Proving for Combinators

How do you tell if two combinator expressions are equal? It depends what you mean by "equal". The simplest definition—that we've implicitly used in constructing multiway graphs—is that expressions are equal only if they're syntactically exactly the same (say they're both s[k][s[s]]).

But what about a more semantic definition, that takes into account the fact that one combinator expression can be transformed to another by the combinator rules? The obvious thing to say is that combinator expressions should be considered equal if they can somehow be transformed by the rules into expressions that are syntactically the same.

And so long as the combinators evolve to fixed points this is in principle straightforward to tell. Like here are four syntactically different combinator expressions that all evolve to the same fixed point, and so in a semantic sense can be considered equal:

s[s][k][k][s[k]]	s[s[s]][s][k][k]	s[s][s][k][s][k]	s[k[s]][k[k]][k]
s[k][k[k]][s[k]]	s[s][k][s[k]][k]	s[k][s[k]][s][k]	k[s][k][k[k][k]]
k[s[k]][k[k][s[k]]]	s[s[k]][k[s[k]]][k]	k[s][s[k]][s]][k]	s[k[k][k]]
s[k]	s[k][k][k[s[k]][k]]	s[k]	s[k]
	k[k[s[k]][k]][k[k[s[k]][k]]]		
	k[s[k]][k]		
	s[k]		

One can think of the fixed point as representing a canonical form to which combinator expressions that are equal can be transformed. One can also think of the steps in the evolution as corresponding to steps in a proof of equality.

But there's already an issue—that's associated with the fundamental fact that combinators are computation universal. Because in general there's no upper bound on how many steps it can take for the evolution of a combinator expression to halt (and no general *a priori* way to even tell if it'll halt at all). So that means that there's also no upper bound on the "length of proof" needed to show by explicit computation that two combinators are equal. Yes, it might only take 12 steps to show that this is yet another combinator equal to s[k]:

s[s[s]][s][s][s][k]

s[s][s][s[s]][s][k]

s[s[s]][s[s[s]]][s][k]

s[s][s][s[s[s]][s]][k]

s[s[s[s]][s]][s[s[s[s]][s]]][k]

s[s[s]][s][k][s[s[s[s]][s]][k]]

s[s][k][s[k]][s[s[s[s]][s]][k]]

s[s[k]][k[s[k]]][s[s[s[s]][s]][k]]

s[k][s[s[s[s]][s]][k]][k[s[k]]][s[s[s[s]][s]][k]]]

k[k[s[k]][s[s[s[s]][s]][k]]][s[s[s[s]][s]][k][k[s[k]][s[s[s[s]][s]][k]]]]

k[s[k]][s[s[s[s]][s]][k]]

s[k]

But it could also take 31 steps (and involve an intermediate expression of size 65):

```
s[s[s]][s][s][s][k]
s[s][s][s[s]][s][k]
s[s[s]][s[s[s]]][s][k]
s[s][s][s[s[s]][s]][k]
s[s[s]][s][s[s[s]][s]][s][k]
s[s[s]][s][s[s[s]][s]][s]][k]
s[s][s][s[s]][s[s[s]][s]][s]][k]
s[s[s]][s[s[s]]][s[s[s]][s]][k]
s[s][s[s[s]][s]][s][s[s[s]][s]][s]][k]
s[s[s]][s[s[s]][s]][s]][s[s][s]][s][s[s[s]][s[s[s]][s]][s]][k]
s[s[s]][s[s[s]][s]][s]][k][s[s[s]][s]][s]][s[s[s]][s[s[s]][s]][s]][k]]
s[s][k][s[s[s]][s]][s][k][s[s[s]][s]][s]][s[s[s]][s[s[s]][s]][s]][k]]
s[s[s[s]][s]][s][k][k[s[s[s]][s]][s]][s][k]][s[s[s]][s]][s][s[s[s]][s[s[s]][s]][s]][k]]
s[s[s]][s]][s][k][s[s[s]][s]][s]][s[s[s]][s[s[s]][s]][s]][k]][k[s[s[s]][s]][s]][s][k]][s[s[s]][s]][s][s[s[s]][s[s[s]][s]][s]][k]]
s[s[s]][s][k][s[k]][s[s[s]][s]][s]][s[s[s]][s[s[s]][s]][s]][k]][k[s[s[s]][s]][s]][s][k]][s[s[s]][s]][s][s[s[s]][s[s[s]][s]][s]][k]]
s[s][k][s[k]][s[k]][s[s[s]][s]][s]][s[s[s]][s[s[s]][s]][s]][k]][k[s[s[s]][s]][s]][s][k]][s[s[s]][s]][s][s[s[s]][s[s[s]][s]][s]][k]]
s[s[k]][k[s[k]]][s[k]][s[s[s]][s]][s]][s[s[s]][s[s[s]][s]][s]][k]][k[s[s[s]][s]][s]][s][k]][s[s[s]][s]][s][s[s[s]][s[s[s]][s]][s]][k]]
s[k][s[s[s[s]][s]][s]][k]][k[s[k]]][s[s[s]][s]][s]][s[s[s]][s[s[s]][s]][s]][k]][k[s[s[s]][s]][s]][s][k]][s[s[s]][s]][s][s[s[s]][s[s[s]][s]][s]][k]]
k[k[s[k]]][s[k]][k[s[k]]][s[k]]]][s[s[s]][s]][s]][s[s[s]][s[s[s]][s]][s]][k]][k[s[s[s]][s]][s]][s][k]][s[s[s]][s]][s][s[s[s]][s[s[s]][s]][s]][k]]
k[s[k]][s[k]]][s[s[s]][s]][s]][s[s[s]][s[s[s]][s]][s]][k]][k[s[s[s]][s]][s]][s][k]][s[s[s]][s]][s][s[s[s]][s[s[s]][s]][s]][k]]
s[k][s[s[s[s]][s]][s]][s[s[s]][s[s[s]][s]][s]][k]][k[s[s[s]][s]][s]][s][k]][s[s[s]][s]][s][s[s[s]][s[s[s]][s]][s]][k]]

k[k[s[s[s]][s]][s]][s][k]][s[s[s]][s]][s]][s[s[s]][s[s[s]][s]][s]][k]][
  s[s[s]][s]][s][s[s[s]][s]][s]][s[s[s]][s[s[s]][s]][s]][k]][k[s[s[s]][s]][s]][s][k]][s[s[s]][s]][s][s[s[s]][s[s[s]][s]][s]][k]]]
k[s[s[s]][s]][s][k]][s[s[s]][s]][s]][s[s[s]][s[s[s]][s]][s]][k]]
s[s[s]][s]][s][k]
s[s[s]][s][k][s[k]]
s[s][k][s[k]][s[k]]
s[s[k]][k[s[k]]][s[k]]
s[k][s[k]][k[s[k]]][s[k]]]
k[k[s[k]]][s[k]]][s[k][k[s[k]]][s[k]]]]
k[s[k]][s[k]]
s[k]
```

We know that if we use leftmost-outermost evaluation, then any combinator expression that has a fixed point will eventually evolve to it (even though we can't in general know how long it will take). But what about combinator expressions that don't have fixed points? How can we tell if they're "equal" according to our definition?

Basically we have to be able to tell if there are sequences of transformations under the combinator rules that cause the expressions to wind up syntactically the same. We can think of these sequences of transformations as being like possible paths of evolution. So then in effect what we're asking is whether there are paths of evolution for different combinators that intersect.

But how can we characterize what possible paths of evolution might exist for all possible evaluation schemes? Well, that's what the multiway graph does. And in terms of multiway graphs there's then a concrete way to ask about equality (or, really, equivalence) between combinator expressions. We basically just need to ask whether there is some appropriate path between the expressions in the multiway graph.

There are lots of details, some of which we'll discuss later. But what we're basically dealing with is a quintessential example of the problem of theorem proving in a formal system. There are different ways to set things up. But as one example, we could take our system to define certain axioms that transform expressions. Applying these axioms in all possible ways generates a multiway graph with expressions as nodes. But then the statement that there's a theorem that expression A is equal to expression B (in the sense that it can be transformed to it) becomes the statement that there's a way to get from A to B in the graph—and giving a path can then be thought of as giving a proof of the theorem.

As an example, consider the combinator expressions:

s[s][s[s][s[s[s]]][k]][k[s[s][s[s[s]]][k]]]

s[k[s[k][s[s[s]][k]]]][k[s[s][s[s[s]]][k]]]

Constructing a multiway graph one can then find a path

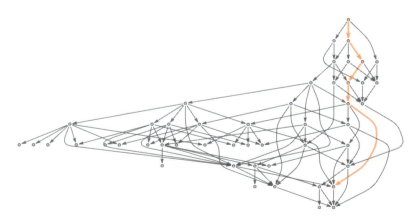

which corresponds to the proof that one can get from one of these expressions
to the other:

```
s[s][s[s][s[s[s]]][k]][k[s[s][s[s[s]]][k]]]
s[k[s[s][s[s[s]]][k]]][s[s][s[s[s]]][k][k[s[s][s[s[s]]][k]]]]
s[k[s[k][s[s[s]][k]]]][s[s][s[s[s]]][k][k[s[s][s[s[s]]][k]]]]
s[k[s[k][s[s[s]][k]]]][s[k][s[s[s]][k]][k[s[s][s[s[s]]][k]]]]
s[k[s[k][s[s[s]][k]]]][k[k[s[s][s[s[s]]][k]]][s[s[s]][k][k[s[s][s[s[s]]][k]]]]]
s[k[s[k][s[s[s]][k]]]][k[s[s][s[s[s]]][k]]]
```

In this particular case, both expressions eventually reach a fixed point. But
consider the expressions:

s[s[s[s][s]]][k]][s[s[s[s][s]]][k]]][k[s[s[s][s]]][k]]]

s[s[s][s]]][k][s[s[s[s][s]]][k]]][k[s[s[s][s]]][k]]]]

Neither of these expressions evolves to a fixed point. But there's still a path in the
(ultimately infinite) multiway graph between them

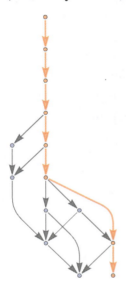

corresponding to the equivalence proof:

```
s[s[s[s][s]]][k]][s[s[s[s][s]]][k]]][k[s[s[s][s]]][k]]]
s[s[s][s]]][k][k[s[s[s][s]]][k]]][s[s[s[s][s]]][k]][k[s[s[s][s]]][k]]]]
s[s[s][s]]][k[s[s[s][s]]][k]]][k[k[s[s[s][s]]][k]]][s[s[s[s][s]]][k]][k[s[s[s][s]]][k]]]]]
s[s][s][k[k[s[s[s][s]]][k]]]][k[s[s[s][s]]][k]][k[k[s[s[s][s]]][k]]]]][s[s[s[s][s]]][k]][k[s[s[s][s]]][k]]]]
s[k[k[s[s[s][s]]][k]]][s[k[k[s[s[s][s]]][k]]]][k[s[s[s][s]]][k]][k[k[s[s[s][s]]][k]]]]]][s[s[s[s][s]]][k]][k[s[s[s][s]]][k]]]]
k[k[s[s[s][s]]][k]]][k[s[s[s][s]]][k]]][k[k[s[s[s][s]]][k]]]]][s[k[k[s[s[s][s]]][k]]]][k[s[s[s][s]]][k]][k[k[s[s[s][s]]][k]]]]]][
   s[s[s[s][s]]][k]][k[s[s[s][s]]][k]]]]
k[s[s[s][s]]][k]][s[k[k[s[s[s][s]]][k]]]][k[s[s[s][s]]][k]][k[k[s[s[s][s]]][k]]]]]]][s[s[s[s][s]]][k]][k[s[s[s][s]]][k]]]]
s[s[s][s]]][k][s[s[s[s][s]]][k]]][k[s[s[s][s]]][k]]]]
```

But with our definition, two combinator expressions can still be considered equal even if one of them can't evolve into the other: it can just be that among the possible ancestors (or, equivalently for combinators, successors) of the expressions there's somewhere an expression in common. (In physics terms, that their light cones somewhere overlap.)

Consider the expressions:

{s[s[s][s]][s][s[s][k]], s[s][k][s[s[s][k]]][k]}

Neither terminates, but it still turns out that there are paths of evolution for each of them that lead to the same expression:

s[s[s][s]][s][s[s][k]]
s[s[s][s[s][k]][s[s[s][k]]
s[s[s][k]][s[s[s][k]]][s[s[s][k]]]

s[s][k][s[s[s][k]]][k]
s[s[s][k]]][k[s[s[s][k]]]][k]
s[s[s][k]][k][k[s[s[s][k]]][k]]
s[s[s][k]][k][k[s[s[s][k]]]
s[s][k][s[s[s][k]]][k[s[s[s][k]]]]
s[s[s][k]]][k[s[s[s][k]]]][k[s[s[s][k]]]]
s[s[s][k]][k[s[s[s][k]]]][k[s[s[s][k]]][k[s[s[s][k]]]]]
s[s][k][k[s[s[s][k]]][k[s[s[s][k]]]]][k[s[s[s][k]]][k[s[s[s][k]]][k[s[s[s][k]]]]]]
s[s][k][k[s[s[s][k]]][k[s[s[s][k]]]]][s[s[s][k]]]
s[s][k][s[s[s][k]]][s[s[s][k]]]
s[s[s][k]]][k[s[s[s][k]]]][s[s[s][k]]]
s[s[s][k]][s[s[s][k]]][k[s[s[s][k]]][s[s[s][k]]]]
s[s][k][s[s[s][k]]][s[s[s][k]]]

If we draw a combined multiway graph starting from the two initial expressions, we can see the converging paths:

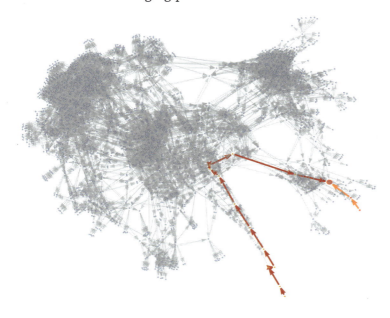

But is there a more systematic way to think about relations between combinator expressions? Combinators are in a sense fundamentally computational constructs. But one can still try to connect them with traditional mathematics, and in particular with abstract algebra.

And so, for example, it's common in the literature of combinators to talk about "combinatory algebra", and to write an expression like

s[k][s[s[k[s[s]][s]][s]][k]][k[s[k]][s[s[k[s[s]][s]][s]][k]]]

as

S•K•(S•(S•(K•(S•S)•S)•S)•K)•(K•(S•K)•(S•(S•(K•(S•S)•S)•S)•K))

where now one imagines that • ("application") is like an algebraic operator that "satisfies the relations"

$$\{\mathbf{S}{\bullet}x{\bullet}y{\bullet}z = x{\bullet}z{\bullet}(y{\bullet}z),\ \mathbf{K}{\bullet}x{\bullet}y = x\}$$

with "constants" **S** and **K**. To determine whether two combinator expressions are equal one then has to see if there's a sequence of "algebraic" transformations that can go from one to the other. The setup is very similar to what we've discussed above, but the "two-way" character of the rules allows one to directly use standard equational logic theorem-proving methods (although because combinator evolution is confluent one never strictly has to use reversed rules).

So, for example, to prove s[k[s]][k[k]][k] = s[s][s][k][s][k] or

S(KS)(KK)K = SSSKSK

one applies a series of transformations based on the **S** and **K** "axioms" to parts of the left- and right-hand sides to eventually reduce the original equation to a tautology:

s•s•s•k •s•k == s•(k•s)•(k•k)•k	s•x•y•z → x•z•(y•z)
s•k•(s•k)•s•k == s•(k•s)•(k•k)•k	s•x•y•z → x•z•(y•z)
s•k•(s•k)•s•k == k•s•k• k•k•k	k•x•y → x
s•k•(s•k)•s •k == k•s•k•k	s•x•y•z → x•z•(y•z)
k•s•(s•k•s)•k == k•s•k •k	k•x•y → x
k•s•(s•k•s)•k == s•k	k•x•y → x
■	True

One can give the outline of this proof as a standard FindEquationalProof proof graph:

The yellowish dots correspond to the "intermediate lemmas" listed above, and the dotted lines indicate which lemmas use which axioms.

One can establish a theorem like

S(KS)(KK)K = S(SS)SSSK

with a slightly more complex proof:

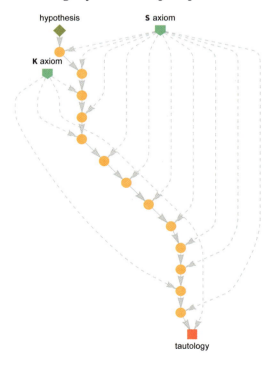

One feature of this proof is that because the combinator rules are confluent—so that different branches in the multiway system always merge—the proof never has to involve critical pair lemmas representing equivalences between branches in the multiway system, and so can consist purely of a sequence of "substitution lemmas".

There's another tricky issue, though. And it has to do with taking "everyday" mathematical notions and connecting them with the precise symbolic structure that defines combinators and their evolution. As an example, let's say you have combinators a and b. It might seem obvious that if a is to be considered equal to b, then it must follow that a[x] = b[x] for all x.

But actually saying this is true is telling us something about what we mean by "equal", and to specify this precisely we have to add the statement as a new axiom.

In our basic setup for proving anything to do with equality (or, for that matter, any equivalence relation), we're already assuming the basic features of equivalence relations (reflexivity, symmetry, transitivity):

$$x = x$$
$$x = y \Rightarrow y = x$$
$$(x = y) \wedge (y = z) \Rightarrow x = z$$

In order to allow us to maintain equality while doing substitutions we also need the axiom:

$$(x = y) \wedge (z = u) \Rightarrow x \cdot z = y \cdot u$$

And now to specify that combinator expressions that are considered equal also "do the same thing" when applied to equal expressions, we need the "extensionality" axiom:

$$x = y \Rightarrow x \cdot z = y \cdot z$$

The previous axioms all work in pure "equational logic". But when we add the extensionality axiom we have to explicitly use full first-order logic—with the result that we get more complicated proofs, though the same basic methods apply.

Lemmas and the Structure of Combinator Space

One feature of the proofs we've seen above is that each intermediate lemma just involves direct use of one or other of the axioms. But in general, lemmas can use lemmas, and one can "recursively" build up a proof much more efficiently than just by always directly using the axioms.

But which lemmas are best to use? If one's doing ordinary human mathematics— and trying to make proofs intended for human consumption—one typically wants to use "famous lemmas" that help create a human-relatable narrative. But realistically there isn't likely to be a "human-relatable narrative" for most combinator equivalence theorems (or, at least there won't be until or unless thinking in terms of combinators somehow becomes commonplace).

So then there's a more "mechanical" criterion: what lemmas do best at reducing the lengths of as many proofs as much as possible? There's some trickiness associated with translations between proofs of equalities and proofs that one expression can evolve into another. But roughly the question boils down to this. When we construct a multiway graph of combinator evolution, each event—and thus each edge—is just the application of a single combinator "axiom".

But if instead we do transformations based on more sophisticated lemmas we can potentially get from one expression to another in fewer steps. In other words, if we "cache" certain combinator transformations, can we make finding paths in combinator multiway graphs systematically more efficient?

To find all possible "combinator theorems" from a multiway system, we should start from all possible combinator expressions, then trace all possible paths to other expressions. It's a little like what we did in the previous section—except now we want to consider multiway evolution with all possible evaluation orders.

Here's the complete multiway graph starting from all size-4 combinator expressions:

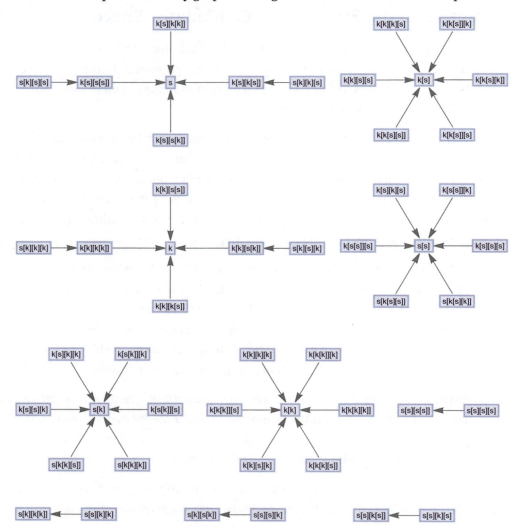

Up to size 6, the graph is still finite (with each disconnected component in effect corresponding to a separate "fixed-point attractor"):

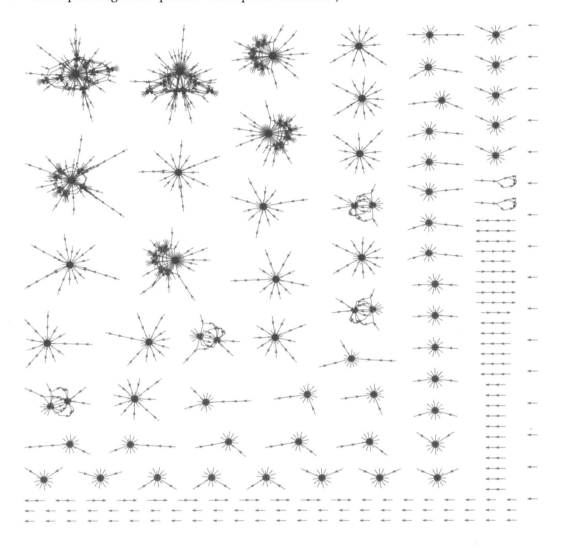

For size 7 and above, it becomes infinite. Here's the beginning of the graph for size-8 expressions involving only s:

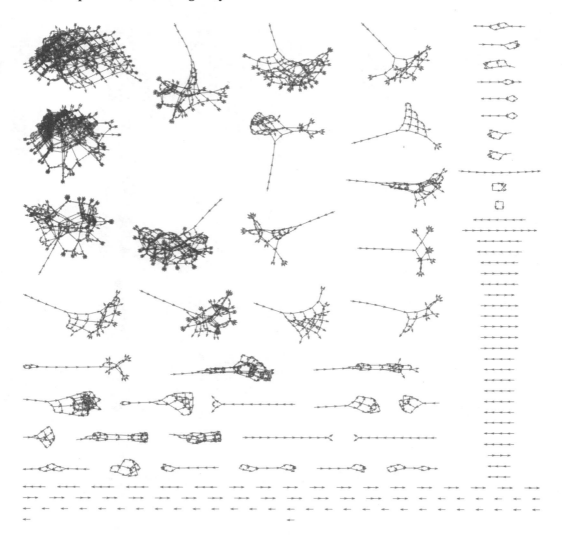

If one keeps only terminating cases, one gets for size 8:

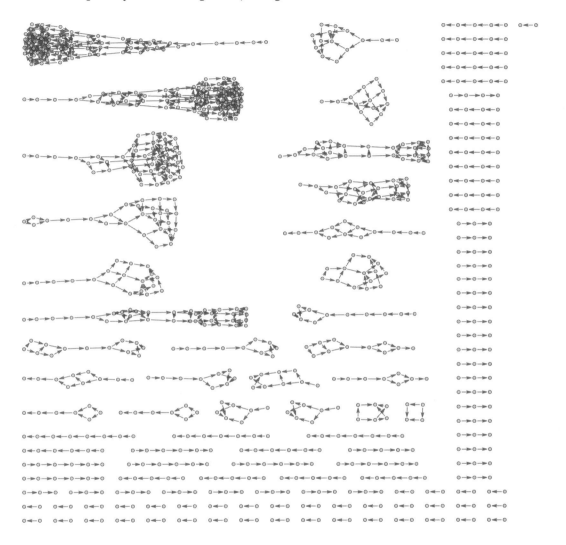

And for size 9:

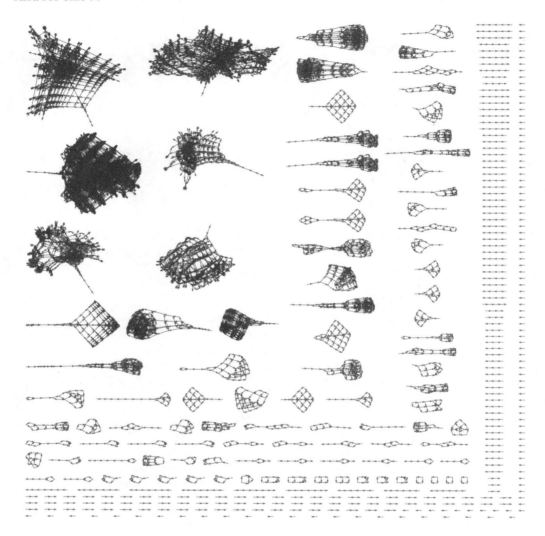

And for size 10:

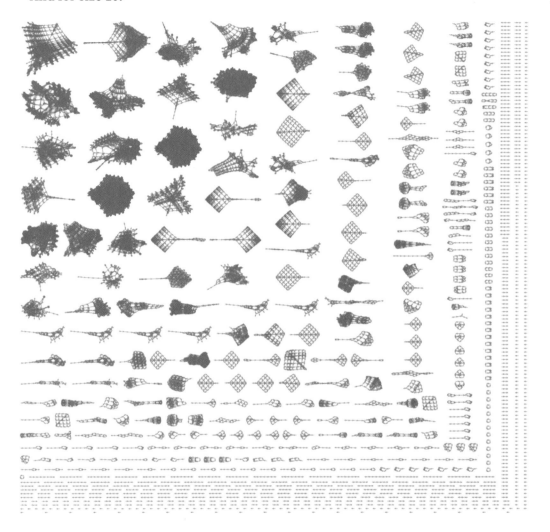

To assess the "most useful" transformations for "finding equations" there's more to do: not only do we need to track what leads to what, but we also need to track causal relationships. And this leads to ideas like using lemmas that have the largest number of causal edges associated with them.

But are there perhaps other ways to find relations between combinator expressions, and combinator theorems? Can we for example figure out what combinator expressions are "close to" what others? In a sense what we need is to define a "space of combinator expressions" with some appropriate notion of nearness.

One approach would just be to look at "raw distances" between trees—say based on asking how many edits have to be made to one tree to get to another.

But an approach that more closely reflects actual features of combinators is to think about the concept of branchial graphs and branchial space that comes from our Physics Project.

Consider for example the multiway graph generated from s[s[s]][s][s[s]][s] (**S(SS)S(SS)S**):

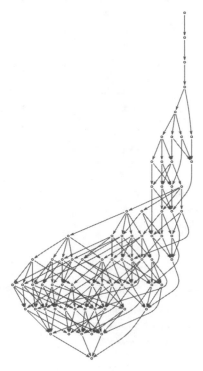

Now consider a foliation of this graph (and in general there will be many possible foliations that respect the partial order defined by the multiway graph):

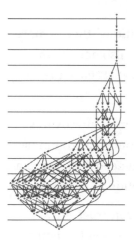

In each slice, we can then define—as in our Physics Project—a branchial graph in which nodes are joined when they have an immediate common ancestor in the multiway graph. In the case shown here, the branchial graphs in successive slices are:

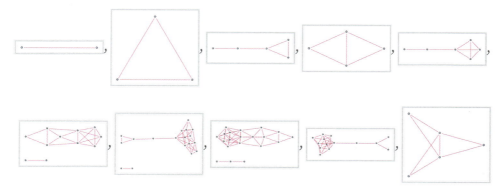

If we consider a combinator expression like s[s][s][s[s]][s][s] (**SSS(SS)SS**) that leads to infinite growth, we can ask what the "long-term" structure of the branchial graph will be. Here are the results after 18 and 19 steps:

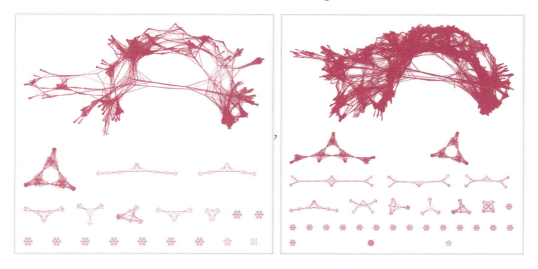

The largest connected components here contain respectively 1879 and 10,693 combinator expressions. But what can we say about their structure? One thing suggested by our Physics Project is to try to "fit them to continuous spaces". And a first step in doing that is to estimate their effective dimension—which one can do by looking at the growth in the volume of a "geodesic ball" in the graph as a function of its radius:

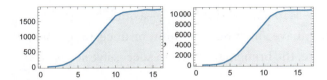

The result for distances small compared to the diameter of the graph is close to quadratic growth—suggesting that there is some sense in which the space of combinator expressions generated in this way may have a limiting 2D manifold structure.

It's worth pointing out that different foliations of the multiway graph (i.e. using different "reference frames") will lead to different branchial graphs—but presumably the (suitably defined) causal invariance of combinator evolution will lead to relativistic-like invariance properties of the branchial graphs.

Somewhat complementary to looking at foliations of the multiway graph is the idea of trying to find quantities that can be computed for combinator expressions to determine whether the combinator expressions can be equal. Can we in essence find hash codes for combinator expressions that are equal whenever the combinator expressions are equal?

In general we've been looking at "purely symbolic" combinator expressions—like:

K•(K•(S•K)•(S•K))•(S•K•(K•(S•K)•(S•K)))

But what if we consider S, K to have definite, say numerical, values, and • to be some kind of generalized multiplication operator that combines these values? We used this kind of approach above in finding a procedure for determining whether S combinator expressions will evolve to fixed points. And in general each possible choice of "multiplication functions" (and S, K "constant values") can be viewed in mathematical terms as setting up a "model" (in the model-theoretic sense) for the "combinatory algebra".

As a simple example, let's consider a finite model in which there are just 2 possible values, and the "multiplication table" for the • operator is:

•	1	2
1	2	1
2	2	2

If we consider S combinator expressions of size 5, there are a total of 14 such expressions, in 10 equivalence classes, that evolve to different fixed points. If we now "evaluate the trees" according to our "model for •" we can see that within each equivalence class the value accumulated at the root of the tree is always the same, but differs between at least some of the equivalence classes:

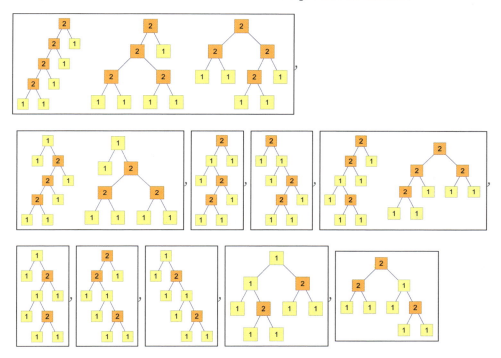

If we look at larger combinator expressions this all keeps working—until we get to two particular size-10 expressions, which have the same fixed point, but different "values":

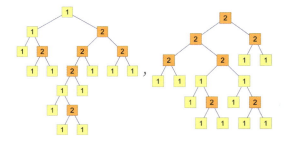

Allowing 3 possible values, the longest-surviving models are

•	1	2	3
1	2	3	2
2	2	2	2
3	2	2	1

,

•	1	2	3
1	3	3	2
2	3	1	3
3	3	3	3

but these both fail at size 13 (e.g. for s[s][s[s]][s[s[s[s]][s][s]]][s[s]]], s[s][s]][s[s[s[s]][s[s[s]]]]]][s[s]]).

The fact that combinator equivalence is in general undecidable means we can't expect to find a computationally finite "valuation procedure" that will distinguish all inequivalent combinator expressions. But it's still conceivable that we could have a scheme to distinguish some classes of combinator expressions from others—in essence through the values of a kind of "conserved quantity for combinators".

Another approach is to consider directly "combinator axioms" like

$$\{S{\bullet}x{\bullet}y{\bullet}z = x{\bullet}z{\bullet}(y{\bullet}z), K{\bullet}x{\bullet}y = x\}$$

and simply ask if there are models of •, S and K that satisfy them. Assuming a finite "multiplication table", there's no way to do this for K, and thus for S and K together. For S alone, however, there are already 8 2-valued models, and 285 3-valued ones.

The full story is more complicated, and has been the subject of a fair amount of academic work on combinators over the past half century. The main result is that there are models that are in principle known to exist, though they're infinite and probably can't be explicitly constructed.

In the case of something like arithmetic, there are formal axioms (the Peano axioms). But we know that (even though Gödel's theorem shows that there are inevitably also other, exotic, non-standard models) there's a model of these axioms that is the ordinary integers. And our familiarity with these and their properties makes us feel that the Peano axioms aren't just formal axioms; they're axioms "about" something, namely integers.

What are the combinator axioms "about"? There's a perfectly good interpretation of them in terms of computational processes. But there doesn't seem to be some "static" set of constructs—like the integers—that give one more insight about what combinators "really are". Instead, it seems, combinators are in the end just through and through computational.

Empirical Computation Theory with Combinators

We've talked a lot here about what combinators "naturally do". But what about getting combinators to do something specific—for example to perform a particular computation we want?

As we saw by example at the beginning of this piece, it's not difficult to take any symbolic structure and "compile it" to combinators. Let's say we're given:

f[y[x]][y][x]

There's then a recursive procedure that effectively builds "function invocations" out of S's and "stops computations" with K's. And using this we can "compile" our symbolic expression to the (slightly complicated) combinator expression:

s[s[k[s]][s[k[s[k[s]]]][s[s[k[s]][s[k[s[k[s]]]]][s[s[k[s]][s[k[k]][s[k[s]][k]]]][k[s[k[s[s[k[k]]]][k]]]]]]][k[k[s[k][k]]]]]]][k[k]]

To "compute our original expression" we just have to take this combinator expression ("■"), form ■[f][x][y], then apply the combinator rules and find the fixed point:

```
s[s[k[s]][s[k[s[k[s]]]]][s[s[k[s]][s[k[s[k[s]]]]][s[s[k[s]][s[k[k]][s[k[s]][k]]]][k[s[k[s[s[k[k]]]][k]]]]]]][k[k[s[k][k]]]]][f][x][y]
s[k[s]][s[k[s[k[s]]]]][s[s[k[s]][s[k[s[k[s]]]]][s[s[k[s]][s[k[k]][s[k[s]][k]]]][k[s[k[s[s[k[k]]]][k]]]]]][k[k[s[k][k]]]][f][k[k]][f][x][y]
k[s][f][s[k[s[k[s]]]][s[s[k[s]][s[k[s[k[s]]]]][s[s[k[s]][s[k[k]][s[k[s]][k]]]][k[s[k[s[s[k[k]]]][k]]]]]][k[k[s[k][k]]]]][f]][k[k][f]][x][y]
s[s[k[s[k[s]]]][s[s[k[s]][s[k[s[k[s]]]]][s[s[k[s]][s[k[k]][s[k[s]][k]]]][k[s[k[s[s[k[k]]]][k]]]]]][k[k[s[k][k]]]]][f][k[k][f]][x][y]
s[k[s[k[s]]]][s[s[k[s]][s[k[s[k[s]]]]][s[s[k[s]][s[k[k]][s[k[s]][k]]]][k[s[k[s[s[k[k]]]][k]]]]]][k[k[s[k][k]]]]][f][x][k[k][f]][x][y]
k[s[k[s]]][f][s[s[k[s]][s[k[s[k[s]]]]][s[s[k[s]][s[k[k]][s[k[s]][k]]]][k[s[k[s[s[k[k]]]][k]]]]]][k[k[s[k][k]]]][f]][x][k[k][f]][x][y]
s[k[s]][s[s[k[s]][s[k[s[k[s]]]]][s[s[k[s]][s[k[k]][s[k[s]][k]]]][k[s[k[s[s[k[k]]]][k]]]]]][k[k[s[k][k]]]][f]][x][k[k][f]][x][y]
k[s][x][s[s[k[s]][s[k[s[k[s]]]]][s[s[k[s]][s[k[k]][s[k[s]][k]]]][k[s[k[s[s[k[k]]]][k]]]]]][k[k[s[k][k]]]][f][x]][k[k][f]][x][y]
s[s[s[k[s]][s[k[s[k[s]]]]][s[s[k[s]][s[k[k]][s[k[s]][k]]]][k[s[k[s[s[k[k]]]][k]]]]]][k[k[s[k][k]]]][f][x]][k[k][f]][x][y]
s[s[k[s]][s[k[s[k[s]]]]][s[s[k[s]][s[k[k]][s[k[s]][k]]]][k[s[k[s[s[k[k]]]][k]]]]]][k[k[s[k][k]]]][f][x][k[k][f]][x][y]
s[k[s]][s[k[s[k[s]]]]][s[s[k[s]][s[k[k]][s[k[s]][k]]]][k[s[k[s[s[k[k]]]][k]]]]][f][k[k[s[k][k]]]][f]][x][k[k][f]][x][y]
k[s][f][s[k[s[k[s]]]][s[s[k[s]][s[k[k]][s[k[s]][k]]]][k[s[k[s[s[k[k]]]][k]]]]][f]][k[k[s[k][k]]]][f]][x][y][k[k][f]][x][y]
s[s[k[s[k[s]]]][s[s[k[s]][s[k[k]][s[k[s]][k]]]][k[s[k[s[s[k[k]]]][k]]]]][f]][k[k[s[k][k]]]][f]][x][k[k][f]][x][y]
s[k[s[k[s]]]][s[s[k[s]][s[k[k]][s[k[s]][k]]]][k[s[k[s[s[k[k]]]][k]]]]][f][k[k[s[k][k]]]][f][x]][k[k][f]][x][y]
k[s[k[s]]][f][s[s[k[s]][s[k[k]][s[k[s]][k]]]][k[s[k[s[s[k[k]]]][k]]]]][f]][k[k[s[k][k]]]][f][x]][k[k][f]][x][y]
s[k[s]][s[s[k[s]][s[k[k]][s[k[s]][k]]]][k[s[k[s[s[k[k]]]][k]]]]][f]][k[k[s[k][k]]]][f][x]][k[k][f]][x][y]
k[s][x][s[s[k[s]][s[k[k]][s[k[s]][k]]]][k[s[k[s[s[k[k]]]][k]]]]][f][x]][k[k[s[k][k]]]][f][x]][k[k][f]][x][y]
s[s[s[k[s]][s[k[k]][s[k[s]][k]]]][k[s[k[s[s[k[k]]]][k]]]]][f][x]][k[k[s[k][k]]]][f][x]][k[k][f]][x][y]
s[s[k[s]][s[k[k]][s[k[s]][k]]]][k[s[k[s[s[k[k]]]][k]]]]][f][x][k[k[s[k][k]]]][f][x]][k[k][f]][x][y]
s[k[s]][s[k[k]][s[k[s]][k]]][f][k[s[k[s[s[k[k]]]][k]]]][f]][x][k[k[s[k][k]]]][f][x]][k[k][f]][x][y]
k[s][f][s[k[k]][s[k[s]][k]][f]][k[s[k[s[s[k[k]]]][k]]]][f]][x][k[k[s[k][k]]]][f][x]][k[k][f]][x][y]
s[s[k[k]][s[k[s]][k]][f]][k[s[k[s[s[k[k]]]][k]]]][f]][x][k[k[s[k][k]]]][f][x]][k[k][f]][x][y]
s[k[k]][s[k[s]][k]][f][k[s[k[s[s[k[k]]]][k]]]][f][x]][k[k[s[k][k]]]][f][x]][k[k][f]][x][y]
k[k][f][s[k[s]][k][f]][k[s[k[s[s[k[k]]]][k]]]][f][x]][k[k[s[k][k]]]][f][x]][k[k][f]][x][y]
k[s[k[s]][k][f]][k[s[k[s[s[k[k]]]][k]]]][f][x]][k[k[s[k][k]]]][f][x]][k[k][f]][x][y]
s[k[s]][k][f][k[s[k[s[s[k[k]]]][k]]]][f][x]][k[k[s[k][k]]]][f][x]][k[k][f]][x][y]
k[s][f][k[f]][k[s[k[s[s[k[k]]]][k]]]][f][x]][k[k[s[k][k]]]][f][x]][k[k][f]][x][y]
s[k[f]][k[s[k[s[s[k[k]]]][k]]]][f][x]][k[k[s[k][k]]]][f][x]][k[k][f]][x][y]
k[f][y][k[s[k[s[s[k[k]]]][k]]]][f][x]][y]][k[k[s[k][k]]]][f][x]][k[k][f]][x][y]
f[k[s[k[s[s[k[k]]]][k]]]][f][x]][y]][k[k[s[k][k]]]][f][x]][y]][k[k][f]][x][y]
f[s[k[s[s[k[k]]]][k]]][x][y]][k[k[s[k][k]]]][f][x]][y]][k[k][f]][x][y]
f[k[s[s[k[k]]]][x]][k[x]][y]][k[k[s[k][k]]]][f][x]][y]][k[k][f]][x][y]
f[s[s[k[k]]][k][x][y]][k[k[s[k][k]]]][f][x]][y]][k[k][f]][x][y]
f[s[k[k]][k][y][k[x]][y]]][k[k[s[k][k]]]][f][x]][y]][k[k][f]][x][y]
f[k[y][k[y]][k[x]][y]]][k[k[s[k][k]]]][f][x]][y]][k[k][f]][x][y]
f[y[k[x]][y]]][k[k[s[k][k]]]][f][x]][y]][k[k][f]][x][y]
f[y[x]][k[s[k][k]][x][y]][k[k][f][x][y]]
f[y[x]][k[s[k][k]][x][y]][k[k][f][x][y]]
f[y[x]][s[k][k][y][k[x][y]]]
f[y[x]][k[y][k[y]]][k[k][f][x][y]]
f[y[x]][y][k[k][f][x][y]]
f[y[x]][y][k[x][y]]
f[y[x]][y][x]
```

But is this the "best combinator way" to compute this result?

There are various different things we could mean by "best". Smallest program? Fastest program? Most memory-efficient program? Or said in terms of combinators: Smallest combinator expression? Smallest number of rule applications? Smallest intermediate expression growth?

In computation theory one often talks theoretically about optimal programs and their characteristics. But when one's used to studying programs "in the wild" one can start to do empirical studies of computation-theoretic questions—as I did, for example, with simple Turing machines in *A New Kind of Science*.

Traditional computation theory tends to focus on asymptotic results about "all possible programs". But in empirical computation theory one's dealing with specific programs—and in practice there's a limit to how many one can look at. But the crucial and surprising fact that comes from studying the computational universe of "programs in the wild" is that actually even very small programs can show highly complex behavior that's in some sense typical of all possible programs. And that means that it's realistic to get intuition—and results—about computation-theoretic questions just by doing empirical investigations of actual, small programs.

So how does this work with combinators? An immediate question to ask is: if one wants a particular expression, what are all the possible combinator expressions that will generate it?

Let's start with a seemingly trivial case: x[x]. With the compilation procedure we used above we get the size-7 combinator expression

s[s[k][k]][s[k][k]]

which (with leftmost-outermost evaluation) generates x[x] in 6 steps:

```
s[s[k][k]][s[k][k]][x]
s[k][k][x][s[k][k][x]]
k[x][k[x]][s[k][k][x]]
x[s[k][k][x]]
x[k[x][k[x]]]
x[x]
```

But what happens if we just start enumerating possible combinator expressions? Up to size 5, none compute x[x]. But at size 6, we have:

```
s[s[s]][s][s[k]][x]
s[s][s[k]][s[s[k]]][x]
s[s[s[k]]][s[k][s[s[k]]]][x]
s[s[k]][x][s[k][s[s[k]]][x]]
s[k][s[k][s[s[k]]][x]][x[s[k][s[s[k]]][x]]]
k[x[s[k][s[s[k]]][x]]][s[k][s[s[k]]][x][x[s[k][s[s[k]]][x]]]]
x[s[k][s[s[k]]][x]]
x[k[x][s[s[k]][x]]]
x[x]
```

So we can "save" one unit of program size, but at the "cost" of taking 9 steps, and having an intermediate expression of size 21.

What if we look at size-7 programs? There are a total of 11 that work (including the one from our "compiler"):

{s[s[s[s]]][s][s[k]], s[s[s]][s[k]][s[k]], s[s][s[k]][s[s[k]]], s[s[s[k]]][s[k][s]], s[s][s[k]][s[k][s]], s[s[s[k]]][s[k][k]], s[s][s[k]][s[k][k]], s[s[k][s]][s[k][s]], s[s[k][s]][s[k][k]], s[s[k][k]][s[k][s]], s[s[k][k]][s[k][k]]}

How do these compare in terms of "time" (i.e. number of steps) and "memory" (i.e. maximum intermediate expression size)? There are 4 distinct programs that all take the same time and memory, there are none that are faster, but there are others that are slowest (the slowest taking 12 steps):

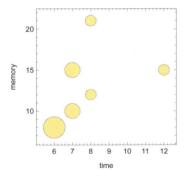

What happens with larger programs? Here's a summary:

size	programs	min time	min memory	max time	max memory	median time	median memory
6	1	9	21	9	21	9	21
7	11	6	8	12	21	7	10
8	95	6	9	18	39	10	15
9	730	6	10	67	207	8	14
10	5754	6	11	375	937	10	15

Here are the distributions of times (dropping outliers)—implying (as the medians above suggest) that even a randomly picked program is likely to be fairly fast:

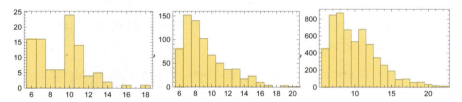

And here's the distribution of time vs. memory on a log-log scale:

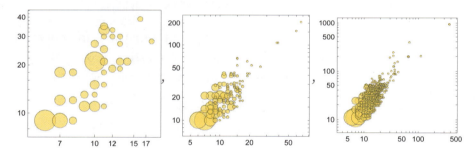

At size 10, the slowest and most memory-intensive program is
s[s[s][k][s[s[s]]]][s][k] (**S(SSK(S(S(SS))))SK**):

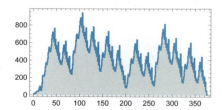

There are so many other questions one can ask. For example: how similar are the various fastest programs? Do they all "work the same way"? At size 7 they pretty much seem to:

At size 8 there are a few "different schemes" that start to appear:

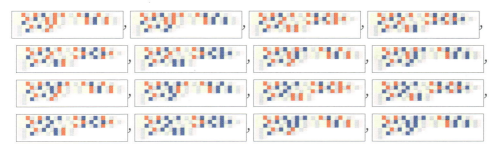

Then one can start to ask questions about how these fastest programs are laid out in the kind of "combinator space" we discussed in the last section—and whether there are good incremental ("evolutionary") ways to find these fastest programs.

Another type of question has to do with the running of our programs. In everything we've done so far in this section, we've used a definite evaluation scheme: leftmost outermost. And in using this definite scheme, we can think of ourselves as doing "deterministic combinator computation". But we can also consider the complete multiway system of all possible updating sequences—which amounts to doing non-deterministic computation.

Here's the multiway graph for the size-6 case we considered above, highlighting the leftmost-outermost evaluation path:

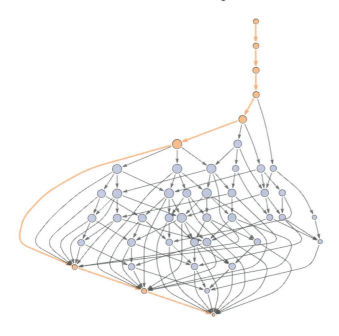

And, yes, in this case leftmost outermost happens to follow a fastest path here. Some other possible schemes are very slow in comparison—with the maximum time being 13 and the maximum intermediate expression size being 21.

At size 7 the multiway graphs for all the leftmost-outermost-fastest programs are the same—and are very simple—among other things making it seem that in retrospect the size-6 case "only just makes it":

At size 8 there are "two ideas" among the 16 cases:

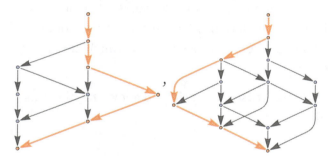

At size 9 there are "5 ideas" among 80 cases:

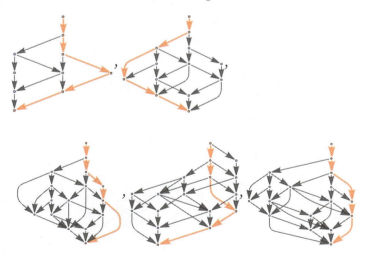

And at size 10 things are starting to get more complicated:

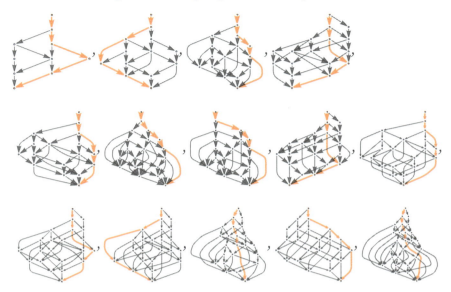

But if we don't look at only leftmost-outermost-fastest programs? At size 7 here are the multiway graphs for all combinator expressions that compute x[x]:

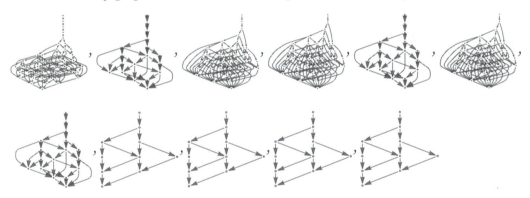

So if one operates "non-deterministically"—i.e. one can follow any path in the multiway graph, not just the leftmost-outermost evaluation scheme one—can one compute the answer faster? The answer in this particular case is no.

But what about at size 8? Of the 95 programs that compute x[x], in most cases the situation is like for size 7 and leftmost outermost gives the fastest result. But there are some wilder things that can happen.

Consider for example:

s[s[s[s]]][k[s[k]]][s]

Here's the complete multiway graph in this case (with 477 nodes altogether):

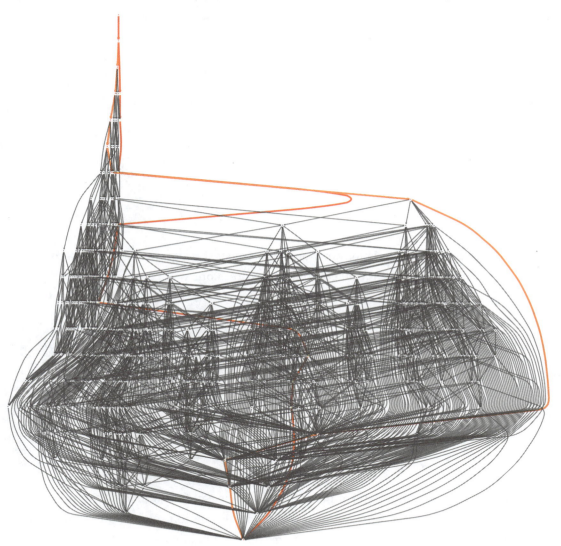

Two paths are indicated: the one in orange is the leftmost-outermost evaluation—
which takes 12 steps in this case. But there's also another path, shown in red—
which has length 11. Here's a comparison:

```
1  s[s[s[s]]][k[s[k]]][s][x]
2  s[s[s]][s][k[s[k]]][s]][x]
3  s[s][k[s[k]][s]][s[k[s[k]][s]]][x]
4  s[s[k[s[k]][s]]][k[s[k]][s][s[k[s[k]][s]]]][x]
5  s[k[s[k]][s]][x][k[s[k]][s][s[k[s[k]][s]]]][x]]
6  k[s[k]][s][k[s[k]][s][s[k[s[k]][s]]][x]][x[k[s[k]][s][s[k[s[k]][s]]][x]]]
7  s[k][k[s[k]][s][s[k[s[k]][s]]][x]][x][x[k[s[k]][s][s[k[s[k]][s]]][x]]]
8  k[x[k[s[k]][s][s[k[s[k]][s]]][x]]][k[s[k]][s][s[k[s[k]][s]]][x][x[k[s[k]][s][s[k[s[k]][s]]][x]]]]
9  x[k[s[k]][s][s[k[s[k]][s]]][x]]
10 x[s[k][s[k[s[k]][s]]][x]]
11 x[k[x][s[k[s[k]][s]][x]]]
12 x[x]
```

```
1  s[s[s[s]]][k[s[k]]][s][x]
2  s[s[s]][s][k[s[k]]][s]][x]
3  s[s[s]][s][s[k]][x]
4  s[s][s[k]][s[s[k]]][x]
5  s[s[s[k]][s[k][s[s[k]]]][x]
6  s[s[k]][x][s[k][s[s[k]]][x]]
7  s[k][s[k][s[s[k]]][x]][x][s[k][s[s[k]]][x]]
8  k[x[s[k][s[s[k]]][x]]][s[k][s[s[k]]][x][x[s[k][s[s[k]]][x]]]]
9  k[x[k[x][s[s[k]]][x]]][s[k][s[s[k]]][x][x[s[k][s[s[k]]][x]]]]
10 k[x[x]][s[k][s[s[k]]][x][x[s[k][s[s[k]]][x]]]]
11 x[x]
```

To get a sense of the "amount of non-determinism" that can occur, we can look at
the number of nodes in successive layers of the multiway graph—essentially the
number of "parallel threads" present at each "non-deterministic step":

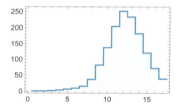

What about size-8 programs for x[x]? There are 9 more—similar to this one—
where the non-deterministic computation is one step shorter. (Sometimes—as
for s[s[s]][s][s[k[k][s]]]—the multiway graph is more complicated, in this case
having 1661 nodes.)

But there are some other things that happen. And a dramatic one is that there can
be paths that just don't terminate at all. s[s[s[s]][s]][s][s[k]] gives an example.
Leftmost-outermost evaluation reaches a fixed point after 14 steps. But overall
the multiway graph grows exponentially (already having size 24,705 after 14
steps)—yielding eventually an infinite number of infinite paths: non-deterministic threads that in a sense get "lost forever".

So far all we've talked about here is the computation of the one—seemingly trivial—
object x[x]. But what about computing other things? Imagine we have a combinator
expression ■ that we apply to x to form ■[x]. If when we "evaluate" this with the
combinator rules it reaches a fixed point we can say this is the result of the computation. But a key point is that most of the time this "result" won't just contain x; it'll
still have "innards of the computation"—in the form of S's and K's—in it.

Out of all 2688 combinator expressions of size 6, 224 compute x. Only one (that we saw above) computes something more complicated: x[x]. At size 7, there are 11 programs that compute x[x], and 4 that compute x[x][x]. At size 8 the things that can be computed are:

x	8926
x[x]	95
x[x[x]]	13
x[x][x]	13
x[x[x]][x]	3
x[x[x]][x[x]][x[x[x[x]]][x[x]]]][x[x[x]][x[x]]]	1

At size 9 the result is:

x	58678
x[x]	730
x[x[x]]	55
x[x][x]	174
x[x[x][x]]	5
x[x][x[x]]	16
x[x[x]][x]	6
x[x][x][x]	3
x[x][x[x[x]]]	9
x[x][x[x]][x]	4
x[x[x]][x[x]]	7
x[x[x]][x]][x]	1
x[x][x][x][x[x][x]]]	1
x[x[x]][x]][x[x][x]]	2
x[x][x[x[x]]]][x[x]]	3
x[x[x]][x[x]][x[x[x]][x[x]]]][x[x[x]][x[x]]]	1
x[x[x][x[x]][x]]][x[x][x[x][x]][x]]][x[x][x[x][x]]]	1

In a sense what we're seeing here are the expressions (or objects) of "low algorithmic information content" with respect to combinator computation: those for which the shortest combinator program that generates them is just of length 9. In addition to shortest program length, we can also ask about expressions generated within certain time or intermediate-expression-size constraints.

What about the other way around? How large a program does one need to generate a certain object? We know that x[x] can be generated with a program of size 6. It turns out x[x[x]] needs a program of size 8:

x[x][x]	s[s[s]][s[s]][s[k]]	7
x[x[x]]	s[s[s[s]]][s][s][s[k]]	8

Here are the shortest programs for objects of size 4:

x[x][x][x]	s[s][s[s[s]]][s[s]][s[k]]	9
x[x[x][x]]	s[s][s[s[s][k]]][s[k][s]]	9
x[x[x]][x]	s[s[s[s][k]]][s[k][s]]	8
x[x[x[x]]]	s[s][s[s[s[s[k]]]]][s[k][s]]	10
x[x][x[x]]	s[s[s[s]]][s[s]][s[k]]	9

Our original "straightforward compiler" generates considerably longer programs: to get an object involving only x's of size n it produces a program of length $4n - 1$ (i.e. 15 in this case).

It's interesting to compare the different situations here. x[x[x]][x[x]][x[x[x][x]][x[x]]]][x[x[x][x]][x[x]]]] (of size 17) can be generated by the program s[s[s]][s][s[s][s[k]]] (of size 8). But the shortest program that can generate x[x[x[x]]] (size 4) is of length 10. And what we're seeing is that different objects can have very different levels of "algorithmic redundancy" under combinator computation.

Clearly we could go on to investigate objects that involve not just x, but also y, etc. And in general there's lots of empirical computation theory that one can expect to do with combinators.

As one last example, one can ask how large a combinator expression is needed to "build to a certain size", in the sense that the combinator expression evolves to a fixed point with that size. Here is the result for all sizes up to 100, both for S, K expressions, and for expressions with S alone (the dotted line is $\log_{\frac{3}{2}}(n)$):

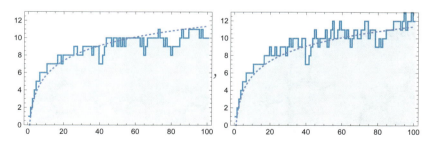

By the way, we can also ask about programs that involve only S, without K. If one wants ■[x] to evaluate to an expression involving only x this isn't possible if one only uses S. But as we discussed above, it's still perfectly possible to imagine "doing a computation" only using S: one just can't expect to have the result delivered directly on its own. Instead, one must run some kind of procedure to extract the result from a "wrapper" that contains S's.

What about practical computations? The most obvious implementation of combinators on standard modern computer systems isn't very efficient because it tends to involve extensive copying of expressions. But by using things like the DAG approach discussed above it's perfectly possible to make it efficient.

What about physical systems? Is there a way to do "intrinsically combinator" computation? As I discussed above, our model of fundamental physics doesn't quite align with combinators. But closer would be computations that can be done with molecules. Imagine a molecule with a certain structure. Now imagine that another molecule reacts with it to produce a molecule with a new structure. If the molecules were tree-like dendrimers, it's at least conceivable that one can get something like a combinator transformation process.

I've been interested for decades in using ideas gleaned from exploring the computational universe to do molecular-scale computation. Combinators as such probably aren't the best "raw material", but understanding how computation works with combinators is likely to be helpful.

And just for fun we can imagine taking actual expressions—say from the evolution of s[s][s][s[s]][s][s]—and converting them to "molecules" just using standard chemical SMILES strings (with C in place of S):

{**SSS(SS)SS, S(SS)(S(SS))SS, SSS(S(SS)S)S, S(S(SS)S)(S(S(SS)S))S,**
  S(SS)SS(S(S(SS)S)S), SSS(SS)(S(S(SS)S)S), S(SS)(S(SS))(S(S(SS)S)S),
  SS(S(S(SS)S)S)(S(SS)(S(S(SS)S)S)), S(S(SS)(S(S(SS)S)S))(S(S(SS)S)S(SS)(S(S(SS)S)S))),
  S(S(SS)(S(S(SS)S)S))(S(SS)S(S(SS)(S(S(SS)S)S))(S(S(SS)(S(S(SS)S)S)))),
  S(S(SS)(S(S(SS)S)S))(SS(S(SS)(S(S(SS)S)S))(S(S(SS)(S(S(SS)S)S)))(S(S(SS)(S(S(SS)S)S))))}

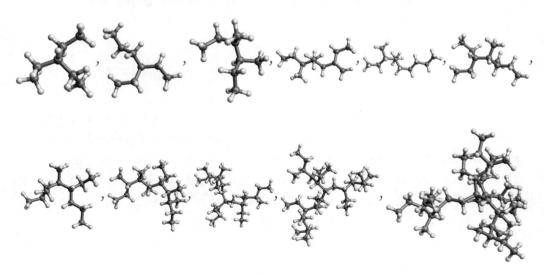

The Future of Combinators

S and K at first seem so simple, so basic. But as we've seen here, there's an immense richness to what they can do. It's a story I've seen played out many times across the computational universe. But in a sense it's particularly remarkable for combinators because they were invented so early, and they seem so very simple.

There's little question that even a century after they were invented, combinators are still hard to get one's head around. Perhaps if computation and computer technology had developed differently, we'd now find combinators easier to understand. Or perhaps the way our brains are made, they're just intrinsically difficult.

In a sense what makes combinators particularly difficult is the extent to which they're both featureless and fundamentally dynamic in their structure. When we apply the ideas of combinators in practical "human-oriented" computing—for example in the Wolfram Language—we annotate what's going on in a variety of ways. But with the Wolfram Physics Project we now have the idea that what happens at the lowest level in the physics of our universe is something much more like "raw combinators".

The details are different—we're dealing with hypergraphs, not trees—but many of the concepts are remarkably similar. Yes, a universe made with combinators probably won't have anything like space in the way we experience it. But a lot of ideas about updating processes and multiway systems are all there in combinators.

For most of their history, combinators have been treated mainly as a kind of backstop for proofs. Yes, it is possible to avoid variables, construct everything symbolically, etc. But a century after they were invented, we can now see that combinators in their own right have much to contribute.

What happens if we don't just think about combinators in general, but actually look at what specific combinators do? What happens if we do experiments on combinators? In the past some elaborate behavior of a particular combinator expression might have just seemed like a curiosity. But now that we have the whole paradigm that I've developed from studying the computational universe we can see how such things fit in, and help build up a coherent story about the ways of computation.

In *A New Kind of Science* I looked a bit at the behavior of combinators; here I've done more. But there's still vastly more to explore in the combinator universe—and many surprises yet to uncover. Doing it will both advance the general science of the computational universe, and will give us a new palette of phenomena and intuition with which to think about other computational systems.

There are things to learn for physics. There are things to learn for language design. There are things to learn about the theoretical foundations of computer science. There may also be things to learn for models of concrete systems in the natural and artificial world—and for the construction of useful technology.

As we look at different kinds of computational systems, several stand out for their minimalism. Particularly notable in the past have been cellular automata, Turing machines and string substitution systems. And now there are also the systems from our Wolfram Physics Project—that seem destined to have all sorts of implications even far beyond physics. And there are also combinators.

One can think of cellular automata, for example, as minimal systems that are intrinsically organized in space and time. The systems from our Wolfram Physics Project are minimal systems that purely capture relations between things. And combinators are in a sense minimal systems that are intrinsically about programs—and whose fundamental structure and operation revolve around the symbolic representation of programs.

What can be done with such things? How should we think about them?

Despite the passage of a century—and a substantial body of academic work—we're still just at the beginning of understanding what can be done with combinators. There's a rich and fertile future ahead, as we begin the second combinator century, now equipped with the ideas of symbolic computational language, the phenomena of the computational universe, and the computational character of fundamental physics.

Historical & Other Notes

I'm writing elsewhere about the origin of combinators, and about their interaction with the history of computation. But here let me make some remarks more specific to this piece.

Combinators were invented in 1920 by Moses Schönfinkel (hence the centenary), and since the late 1920s there's been continuous academic work on them—notably over more than half a century by Haskell Curry.

A classic summary of combinators from a mathematical point of view is the book: Haskell B. Curry and Robert Feys, *Combinatory Logic* (1958). More recent treatments (also of lambda calculus) include: H. P. Barendregt, *The Lambda Calculus* (1981) and J. Roger Hindley and Jonathan P. Seldin, *Lambda-Calculus and Combinators* (1986).

In the combinator literature, what I call "combinator expressions" are often called "terms" (as in "term rewriting systems"). The part of the expression that gets rewritten is often called the "redex"; the parts that get left over are sometimes called the "residuals". The fixed point to which a combinator expression evolves is often called its "normal form", and expressions that reach fixed points are called "normalizing".

Forms like a[b[a][c]] that I "immediately apply to arguments" are basically lambda expressions, written in Wolfram Language using Function. The procedure of "compiling" from lambda expressions to combinators is sometimes called bracket abstraction. As indicated by examples at the end of this piece, there are many possible methods for doing this.

The scheme for doing arithmetic with combinators at the beginning of this piece is based on work by Alonzo Church in the 1930s, and uses so-called "Church numerals". The idea of encoding logic by combinators was discussed by Schönfinkel in his original paper, though the specific minimal encoding I give was something I found by explicit computational search in just the past few weeks. Note that if one uses s[k] for True and k for False (as in the rule 110 cellular automaton encoding) the minimal forms for the Boolean operators are:

15		True	k[k[s[k]]]	K(K(SK))
14		Or	s[s][k]	SSK
13			s[k[s[s[s[s[k]]][s]]]][k]	S(K(S(S(S(SK))S)))K
12		First	k	K
11		Implies	s[s[s[k]]][s]	S(S(SK))S
10		Last	s[k]	SK
9		Equal	s[s[s[s]][s[s[s[k]]]][s]]][k]	S(S(SS)(S(S(SK)))S)K
8		And	s[s[s]][s][s[k]]	S(SS)S(SK)
7		Nand	s[s[s[s][k[k[k[k]]]]]][k[s]]	S(S(SS(K(K(KK))))(KS))
6		Xor	s[s][k[s[s][k[k[k]]]][s]]	SS(K(S(SS(K(KK)))S))
5		Not	k[s[s[s][k[k[k]]]][s]]	K(S(SS(K(KK)))S)
4			s[k[s[s[s][k[k[k]]]]]][k]	S(K(S(SS(K(KK)))))K
3		Not	s[s][s[s[s[s[k]]][s]]][k[k]]	SS(S(S(SK))S)(KK)
2			s[s][k[k[k]]]	SS(K(KK))
1		Nor	s[k[s[s][k[k[k]]]]]][s]	S(S(K(S(SS(K(KK))))))S
0		False	k[k[k]]	K(KK)

The uniqueness of the fixed point for combinators is a consequence of the Church–Rosser property for combinators from 1941. It is closely related to the causal invariance property that appears in our model of physics.

There's been a steady stream of specific combinators defined for particular mathematical purposes. An example is the Y combinator s[s][k][s[k[s[s][s[s[s][k]]]]][k]], which has the property that for any x, Y[x] can be proved to be equivalent to x[Y[x]], and "recurses forever". Here's how Y[x] grows if one just runs it with leftmost-outermost evaluation (and it produces expressions of the form Nest[x, _, n] at step $n^2 + 7n$):

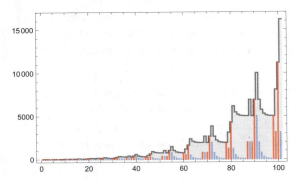

The Y combinator was notably used by Paul Graham in 2005 to name his Y Combinator startup accelerator. And perhaps channeling the aspirations of startups the "actual" Y combinator goes through many ups and downs but (with leftmost-outermost evaluation) reaches size 1 billion ("unicorn") after 494 steps—and after 1284 steps reaches more-dollars-than-in-the-world size: 508,107,499,710,983.

Empirical studies of the actual behavior of combinators "in the wild" have been pretty sparse. The vast majority of academic work on combinators has been done by hand, and without the overall framework of *A New Kind of Science* the detailed behavior of actual combinators mostly just seemed like a curiosity.

I did fairly extensive computational exploration of combinators (and in general what I called "symbolic systems") in the 1990s for *A New Kind of Science*. Page 712 summarized some combinator behavior I found (with /. evaluation):

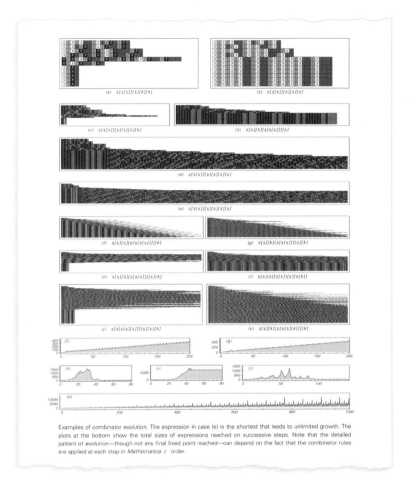

Examples of combinator evolution. The expression in case (e) is the shortest that leads to unlimited growth. The plots at the bottom show the total sizes of expressions reached on successive steps. Note that the detailed pattern of evolution—though not any final fixed point reached—can depend on the fact that the combinator rules are applied at each step in *Mathematica* /. order.

I don't know to what extent the combinator results in *A New Kind of Science* were anticipated elsewhere. Longtime combinator enthusiast Henk Barendregt for example recently pointed me to a paper of his from 1976 mentioning nontermination in S combinator expressions:

The length of an S-term is the number of its S's. If a_n is the number of S-terms with length n, then by the formula of Catalan (cf. [3] p. 64)

$$a_n = \frac{1}{2n-1} \binom{2n-1}{n}.$$

The first values of a_n are indicated in fig. 1.
Let b_n be the number of S-terms of length n without a nf.
Mr. Duboué has calculated by computer upper bounds for b_n, for n < 10, see fig. 1.

n	1	2	3	4	5	6	7	8	9	10
a_n	1	1	2	5	14	42	132	429	1430	4862
b_n	0	0	0	0	0	0	2	≤39	≤231	

(fig. 1)

The bounds are not exact, since the computer only reduced a term a (large) finite number of times in order to conclude that it might be non-normal. For n = 7, theorem 6.4 proves that the bound is exact.

7.1. Notations. C[ ] is a context containing one or more holes.
$F^0 X = X$; $F^{n+1} = F(F^n X)$.
$M \xrightarrow{\oplus} N \iff CL \vdash M \longrightarrow C[N]$ for some context C[ ], and $M \neq C N$.

The procedure I describe for determining the termination of S combinator expression was invented by Johannes Waldmann at the end of the 1990s. (The detailed version that I used here came from Jörg Endrullis.)

What we call multiway systems have been studied in different ways in different fields, under different names. In the case of combinators, they are basically Böhm trees (named after Corrado Böhm).

I've concentrated here on the original S, K combinators; in recent livestreams, as in *A New Kind of Science*, I've also been exploring other combinator rules.

Thanks

Matthew Szudzik has helped me with combinator matters since 1998 (and has given a lecture on combinators almost every year for the past 18 years at our Wolfram Summer School). Roman Maeder did a demo implementation of combinators in Mathematica in 1988, and has now added CombinatorS to Version 12.2 of Wolfram Language.

I've had specific help on this piece from Jonathan Gorard, Jose Martin-Garcia, Eric Paul, Ed Pegg, Max Piskunov, and particularly Mano Namuduri, as well as Jeremy Davis, Sushma Kini, Amy Simpson and Jessica Wong. We've had recent interactions about combinators with a four-academic-generation sequence of combinator researchers: Henk Barendregt, Jan Willem Klop, Jörg Endrullis and Roy Overbeek. Thanks also to John Tromp for suggesting some combinator optimizations after this was first published.

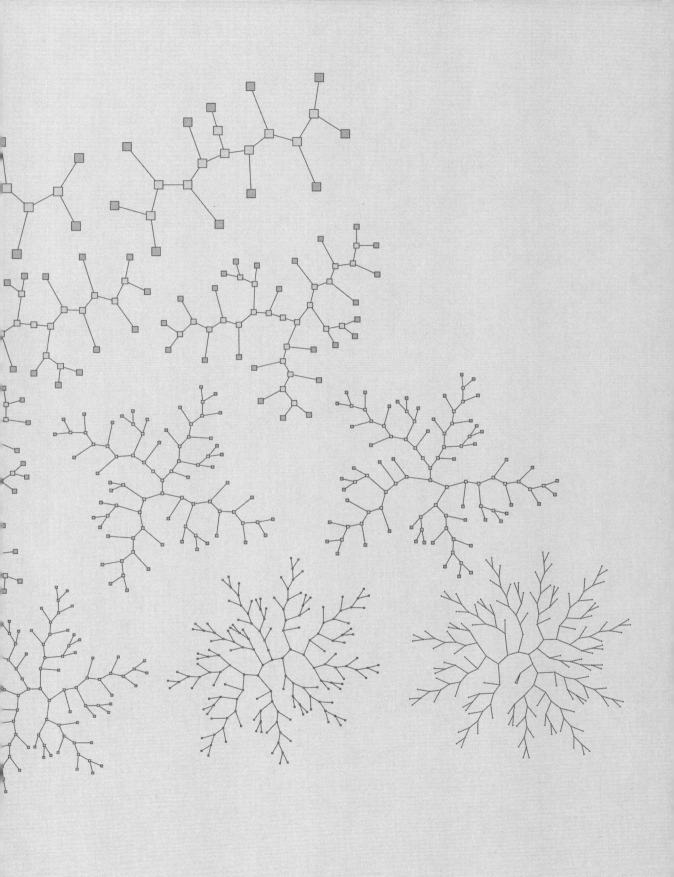

Combinators and the Story of Computation

The Abstract Representation of Things

"In principle you could use combinators," some footnote might say. But the implication tends to be "But you probably don't want to." And, yes, combinators are deeply abstract—and in many ways hard to understand. But tracing their history over the hundred years since they were invented, I've come to realize just how critical they've actually been to the development of our modern conception of computation—and indeed my own contributions to it.

The idea of representing things in a formal, symbolic way has a long history. In antiquity there was Aristotle's logic and Euclid's geometry. By the 1400s there was algebra, and in the 1840s Boolean algebra. Each of these was a formal system that allowed one to make deductions purely within the system. But each, in a sense, ultimately viewed itself as being set up to model something specific. Logic was for modeling the structure of arguments, Euclid's geometry the properties of space, algebra the properties of numbers; Boolean algebra aspired to model the "laws of thought".

But was there perhaps some more general and fundamental infrastructure: some kind of abstract system that could ultimately model or represent anything? Today we understand that's what computation is. And it's becoming clear that the modern conception of computation is one of the single most powerful ideas in all of intellectual history—whose implications are only just beginning to unfold.

But how did we finally get to it? Combinators had an important role to play, woven into a complex tapestry of ideas stretching across more than a century.

The main part of the story begins in the 1800s. Through the course of the 1700s and 1800s mathematics had developed a more and more elaborate formal structure that seemed to be reaching ever further. But what really was mathematics? Was it a formal way of describing the world, or was it something else—perhaps something that could exist without any reference to the world?

Developments like non-Euclidean geometry, group theory and transfinite numbers made it seem as if meaningful mathematics could indeed be done just by positing abstract axioms from scratch and then following a process of deduction. But could all of mathematics actually just be a story of deduction, perhaps even ultimately derivable from something seemingly lower level—like logic?

But if so, what would things like numbers and arithmetic be? Somehow they would have to be "constructed out of pure logic". Today we would recognize these efforts as "writing programs" for numbers and arithmetic in a "machine code" based on certain "instructions of logic". But back then, everything about this and the ideas around it had to be invented.

What Is Mathematics—and Logic—Made Of?

Before one could really dig into the idea of "building mathematics from logic" one had to have ways to "write mathematics" and "write logic". At first, everything was just words and ordinary language. But by the end of the 1600s mathematical notation like +, =, > had been established. For a while new concepts—like Boolean algebra —tended to just piggyback on existing notation. By the end of the 1800s, however, there was a clear need to extend and generalize how one wrote mathematics.

In addition to algebraic variables like x, there was the notion of symbolic functions f, as in $f(x)$. In logic, there had long been the idea of letters ($p, q, ...$) standing for propositions ("it is raining now"). But now there needed to be notation for quantifiers ("for all x such-and-such", or "there exists x such that..."). In addition, in analogy to symbolic functions in mathematics, there were symbolic logical predicates: not just explicit statements like $x > y$ but also ones like $p(x, y)$ for symbolic p.

The first full effort to set up the necessary notation and come up with an actual scheme for constructing arithmetic from logic was Gottlob Frege's 1879 *Begriffsschrift* ("concept script"):

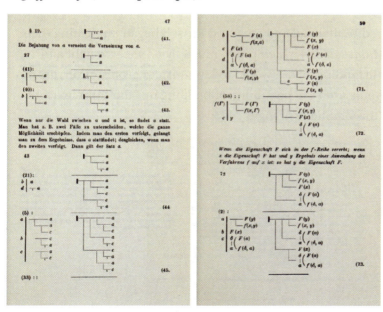

And, yes, it was not so easy to read, or to typeset—and at first it didn't make much of an impression. But the notation got more streamlined with Giuseppe Peano's *Formulario* project in the 1890s—which wasn't so concerned with starting from logic as starting from some specified set of axioms (the "Peano axioms"):

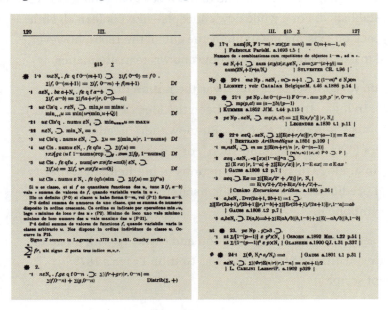

And then in 1910 Alfred Whitehead and Bertrand Russell began publishing their 2000-page *Principia Mathematica*—which pretty much by its sheer weight and ambition (and notwithstanding what I would today consider grotesque errors of language design)—popularized the possibility of building up "the complexity of mathematics" from "the simplicity of logic":

It was one thing to try to represent the content of mathematics, but there was also the question of representing the infrastructure and processes of mathematics. Let's say one picks some axioms. How can one know if they're consistent? What's involved in proving everything one can prove from them?

In the 1890s David Hilbert began to develop ideas about this, particularly in the context of tightening up the formalism of Euclid's geometry and its axioms. And after *Principia Mathematica*, Hilbert turned more seriously to the use of logic-based ideas to develop "metamathematics"—notably leading to the formulation of things like the "decision problem" (*Entscheidungsproblem*) of asking whether, given an axiom system, there's a definite procedure to prove or disprove any statement with respect to it.

But while connections between logic and mathematics were of great interest to people concerned with the philosophy of mathematics, a more obviously mathematical development was universal algebra—in which axioms for different areas of mathematics were specified just by giving appropriate algebraic-like relations. (As it happens, universal algebra was launched under that name by the 1898 book *A Treatise on Universal Algebra* by Alfred Whitehead, later of *Principia Mathematica* fame.)

But there was one area where ideas about algebra and logic intersected: the tightening up of Boolean algebra, and in particular the finding of simpler foundations for it. Logic had pretty much always been formulated in terms of AND, OR and NOT. But in 1912 Henry Sheffer—attempting to simplify *Principia Mathematica*—showed that just NAND (or NOR) were sufficient. (It turned out that Charles Peirce had already noted the same thing in the 1880s.)

So that established that the notation of logic could be made basically as simple as one could imagine. But what about its actual structure, and axioms? Sheffer talked about needing five "algebra-style" axioms. But by going to axioms based on logical inferences Jean Nicod managed in 1917 to get it down to just one axiom. (And, as it happens, I finally finished the job in 2000 by finding the very simplest "algebra-style" axioms for logic—the single axiom: $((p \cdot q) \cdot r) \cdot (p \cdot ((p \cdot r) \cdot p) = r$.)

The big question had in a sense been "What is mathematics ultimately made of?". Well, now it was known that ordinary propositional logic could be built up from very simple elements. So what about the other things used in mathematics—like functions and predicates? Was there a simple way of building these up too?

People like Frege, Whitehead and Russell had all been concerned with constructing specific things—like sets or numbers—that would have immediate mathematical meaning. But Hilbert's work in the late 1910s began to highlight the idea of looking instead at metamathematics and the "mechanism of mathematics"—and in effect at how the pure symbolic infrastructure of mathematics fits together (through proofs, etc.), independent of any immediate "external" mathematical meaning.

Much as Aristotle and subsequent logicians had used (propositional) logic to define a "symbolic structure" for arguments, independent of their subject matter, so too did Hilbert's program imagine a general "symbolic structure" for mathematics, independent of particular mathematical subject matter.

And this is what finally set the stage for the invention of combinators.

Combinators Arrive

We don't know how long it took Moses Schönfinkel to come up with combinators. From what we know of his personal history, it could have been as long as a decade. But it could also have been as short as a few weeks.

There's no advanced math or advanced logic involved in defining combinators. But to drill through the layers of technical detail of mathematical logic to realize that it's even conceivable that everything can be defined in terms of them is a supreme achievement of a kind of abstract reductionism.

There is much we don't know about Schönfinkel as a person. But the 11-page paper he wrote on the basis of his December 7, 1920, talk in which he introduced combinators is extremely clear.

The paper is entitled "On the Building Blocks of Mathematical Logic" (in the original German, "Über die Bausteine der mathematischen Logik".) In other words, its goal is to talk about "atoms" from which mathematical logic can be built. Schönfinkel explains that it's "in the spirit of" Hilbert's axiomatic method to build everything from as few notions as possible; then he says that what he wants to do is to "seek out those notions from which we shall best be able to construct all other notions of the branch of science in question".

His first step is to explain that Hilbert, Whitehead, Russell and Frege all set up mathematical logic in terms of standard AND, OR, NOT, etc. connectives—but that Sheffer had recently been able to show that just a single connective (indicated by a stroke "|"—and what we would now call NAND) was sufficient:

> a und b ist mindestens eine falsch", die mit den obigen Zeichen in den beiden äquivalenten Formen
> $$\bar{a} \vee \bar{b} \quad \text{und} \quad \overline{a \,\&\, b}$$
> geschrieben werden kann, und als zugehöriges neues Zeichen
> $$a \mid b,$$
> so ist augenscheinlich
> $$\bar{a} = a \mid a, \quad a \vee b = (a \mid a) \mid (b \mid b),$$
> womit wegen
> $$a \,\&\, b = \overline{\bar{a} \vee \bar{b}}, \quad (a \rightarrow b) = \bar{a} \vee b, \quad (a \sim b) = (a \rightarrow b) \,\&\, (b \rightarrow a)$$

But in addition to the "content" of these relations, I think Schönfinkel was trying to communicate by example something else: that all these logical connectives can ultimately be thought of just as examples of "abstract symbolic structures" with a certain "function of arguments" (i.e. $f[x,y]$) form.

The next couple of paragraphs talk about how the quantifiers "for all" ($\forall$) and "there exists" ($\exists$) can also be simplified in terms of the Sheffer stroke (i.e. NAND).

But then comes the rallying cry: "The successes that we have encountered thus far... encourage us to attempt further progress." And then he's ready for the big idea—which he explains "at first glance certainly appears extremely bold". He proposes to "eliminate by suitable reduction the remaining fundamental concepts of proposition, function and variable".

He explains that this only makes sense for "arbitrary, logically general propositions", or, as we'd say now, for purely symbolic constructs without specific meanings yet assigned. In other words, his goal is to create a general framework for operating on arbitrary symbolic expressions independent of their interpretation.

He explains that this is valuable both from a "methodological point of view" in achieving "the greatest possible conceptual uniformity", but also from a certain philosophical or perhaps aesthetic point of view.

And in a sense what he was explaining—back in 1920—was something that's been a core part of the computational language design that I've done for the past 40 years: that everything can be represented as a symbolic expression, and that there's tremendous value to this kind of uniformity.

But as a "language designer" Schönfinkel was an ultimate minimalist. He wanted to get rid of as many notions as possible—and in particular he didn't want variables, which he explained were "nothing but tokens that characterize certain argument places and operators as belonging together"; "mere auxiliary notions".

Today we have all sorts of mathematical notation that's at least somewhat "variable free" (think coordinate-free notation, category theory, etc.) But in 1920 mathematics as it was written was full of variables. And it needed a serious idea to see how to get rid of them. And that's where Schönfinkel starts to go "even more symbolic".

He explains that he's going to make a kind of "functional calculus" (*Funktionalkalkül*). He says that normally functions just define a certain correspondence between the domain of their arguments, and the domain of their values. But he says he's going to generalize that—and allow ("disembodied") functions to appear as arguments and values of functions. In other words, he's inventing what we'd now call higher-order functions, where functions can operate "symbolically" on other functions.

In the context of traditional calculus-and-algebra-style mathematics it's a bizarre idea. But really it's an idea about computation and computational structures— that's more abstract and ultimately much more general than the mathematical objectives that inspired it.

But back to Schönfinkel's paper. His next step is to explain that once functions can have other functions as arguments, functions only ever need to take a single argument. In modern (Wolfram Language) notation he says that you never need f[x,y]; you can always do everything with f[x][y].

In something of a sleight of hand, he sets up his notation so that *fxyz* (which might look like a function of three arguments f[x,y,z]) actually means (((*fx*)*y*)*z*) (i.e. f[x][y][z]). (In other words—somewhat confusingly with respect to modern standard functional notation—he takes function application to be left associative.)

Again, it's a bizarre idea—though actually Frege had had a similar idea many years earlier (and now the idea is usually called currying, after Haskell Curry, who we'll be talking about later). But with his "functional calculus" set up, and all functions needing to take only one argument, Schönfinkel is ready for his big result.

He's effectively going to argue that by combining a small set of particular functions he can construct any possible symbolic function—or at least anything needed for predicate logic. He calls them a "sequence of particular functions of a very general nature". Initially there are five of them: the identity function (*Identitätsfunktion*) *I*, the constancy function (*Konstanzfunktion*) *C* (which we now call *K*), the interchange function (*Vertauschungsfunktion*) *T*, the composition function (*Zusammensetzungsfunktion*) *Z*, and the fusion function (*Verschmelzungsfunktion*) *S*.

§ 3.

Es soll nunmehr eine Reihe von *individuellen Funktionen* von sehr allgemeiner Natur eingeführt werden. Ich nenne sie: die Identitätsfunktion *I*, die Konstanzfunktion *C*, die Vertauschungsfunktion *T*, die Zusammensetzungsfunktion *Z* und die Verschmelzungsfunktion *S*.

1. Unter der *Identitätsfunktion I* soll diejenige völlig bestimmte Funktion verstanden werden, deren Argumentwert keiner Einschränkung unterworfen ist und deren Funktionswert stets mit dem Argumentwert übereinstimmt, durch die also jedes Ding und jede Funktion sich selbst zugeordnet wird. Sie ist somit definiert durch die Gleichung

$$I\,x = x,$$

in welcher das Gleichheitszeichen nicht etwa als logische Äquivalenz im Sinne der im logischen Aussagenkalkül üblichen Definition zu lesen ist, sondern besagt, daß die Ausdrücke links und rechts dasselbe bedeuten, d. h. daß der Funktionswert *I x* stets derselbe ist wie der Argumentwert *x*, was man auch für *x* einsetzen mag. (So wäre z. B. *I I = I.*)

2. Nunmehr sei der Argumentwert wieder ohne Einschränkung beliebig, während der Funktionswert unabhängig von jenem stets der feste Wert *a* sein soll. Diese Funktion ist ihrerseits von *a* abhängig, also von der Form *C a*. Daß ihr Funktionswert stets *a* ist, wird geschrieben:

$$(C\,a)\,y = a.$$

Und, indem wir nun auch *a* variabel lassen, erhalten wir:

$$(C\,x)\,y = x \quad \text{bzw.} \quad C\,x\,y = x$$

als Definitionsgleichung der *Konstanzfunktion C*. Diese Funktion *C* ist augenscheinlich von der auf S. 308 betrachteten Art; sie liefert nämlich erst durch Einsetzen eines festen Wertes für *x* eine Funktion mit dem Argument *y*. In der praktischen Anwendung leistet sie uns den Dienst, daß sie eine Größe *x* als „blinde" Veränderliche einzuführen gestattet.

And then he's off and running defining what we now call combinators. The definitions look simple and direct. But to get to them Schönfinkel effectively had to cut away all sorts of conceptual baggage that had come with the historical development of logic and mathematics.

Even talking about the identity combinator isn't completely straightforward. Schönfinkel carefully explains that in $I\,x = x$, equality is direct symbolic or structural equality, or as he puts it "the equal sign is not to be taken to represent logical equivalence as it is ordinarily defined in the propositional calculus of logic but signifies that the expressions on the left and on the right mean the same thing, that is, that the function value lx is always the same as the argument value x, whatever we may substitute for x." He then adds parenthetically, "Thus, for instance, $I\,I$ would be equal to I". And, yes, to someone used to the mathematical idea that a function takes values like numbers, and gives back numbers, this is a bit mind-blowing.

Next he explains the constancy combinator, that he called C (even though the German word for it starts with K), and that we now call K. He says "let us assume that the argument value is again arbitrary without restriction, while, regardless of what this value is, the function value will always be the fixed value a". And when he says "arbitrary" he really means it: it's not just a number or something; it's what we would now think of as any symbolic expression.

First he writes $(C\,a)y = a$, i.e. the value of the "constancy function $C\,a$ operating on any y is a", then he says to "let a be variable too", and defines $(C\,x)y = x$ or $Cxy = x$. Helpfully, almost as if he were writing computer documentation, he adds: "In practical applications C serves to permit the introduction of a quantity x as a 'blind' variable."

Then he's on to T. In modern notation the definition is $T[f][x][y] = f[y][x]$ (i.e. T is essentially ReverseApplied). (He wrote the definition as $(T\phi)\,xy = \phi yx$, explaining that the parentheses can be omitted.) He justifies the idea of T by saying that "The function T makes it possible to alter the order of the terms of an expression, and in this way it compensates to a certain extent for the lack of a commutative law."

Next comes the composition combinator Z. He explains that "In [mathematical] analysis, as is well known, we speak loosely of a 'function of a function'...", by which he meant that it was pretty common then (and now) to write something like $f(g(x))$. But then he "went symbolic"—and defined a composition function that could symbolically act on any two functions f and g: $Z[f][g][x] = f[g[x]]$. He explains that Z allows one to "shift parentheses" in an expression: i.e. whatever the objects in an expression might be, Z allows one to transform [][][] to [[]] etc. But in case this might have seemed too abstract and symbolic, he then attempted to explain in a more "algebraic" way that the effect of Z is "somewhat like that of the associative law" (though, he added, the actual associative law is not satisfied).

Finally comes the *pièce de résistance*: the S combinator (that Schönfinkel calls the "fusion function"):

> 5. Setzt man in
> $$f\,x\,y$$
> für y den Wert einer Funktion g ein, und zwar genommen für dasselbe x, das als Argument von f auftritt, so kommt man auf einen Ausdruck
> $$f\,x\,(g\,x)$$
> oder, wie wir für den Augenblick etwas übersichtlicher schreiben wollen:
> $$(f\,x)(g\,x).$$
> Dies ist natürlich der Wert einer Funktion von x allein, also
> $$(f\,x)(g\,x) = F\,x,$$
> wo
> $$F = S'(f, g)$$
> wieder in einer völlig bestimmten Weise von den gegebenen Funktionen f und g abhängt. Wir haben demgemäß:
> $$[S'(\varphi, \chi)]\,x = (\varphi\,x)(\chi\,x)$$
> oder, nach der auch im vorigen Fall verwendeten Umformung:
> $$S\,\varphi\,\chi\,x = (\varphi\,x)(\chi\,x)$$
> als Definitionsgleichung der *Verschmelzungsfunktion S*.
>
> Es wird gut sein, diese Funktion durch ein praktisches Beispiel dem Verständnis näherzubringen. Nehmen wir etwa für $f\,x\,y$ den Wert $^x\log y$ (d. h. den Logarithmus von y zu der Basis x) und für $g\,z$ den Funktionswert $1 + z$, so ergibt sich $(f\,x)(g\,x)$ augenscheinlich als $^x\log(1 + x)$, d. h. als der Wert einer Funktion von x, die mit den beiden gegebenen Funktionen eben durch unsere allgemeine Funktion S eindeutig verknüpft ist.
>
> Der praktische Nutzen der Funktion S besteht ersichtlich darin, daß sie es ermöglicht, mehrmals auftretende Veränderliche — und bis zu einem gewissen Grade auch individuelle Funktionen — nur einmal auftreten zu lassen.

He doesn't take too long to define it. He basically says: consider $(fx)\,(gx)$ (i.e. f[x][g[x]]). This is really just "a function of x". But what function? It's not a composition of f and g; he calls it a "fusion", and he defines the S combinator to create it: $S[f][g][x] = f[x][g[x]]$.

It's pretty clear Schönfinkel knew this kind of "symbolic gymnastics" would be hard for people to understand. He continues: "It will be advisable to make this function more intelligible by means of a practical example." He says to take fxy (i.e. f[x][y]) to be $\log_x y$ (i.e. Log[x,y]), and gz (i.e. g[z]) to be $1 + z$. Then $Sfgx = (fx)\,(gx) = \log_x(1+x)$ (i.e. S[f][g][x]=f[x][g[x]]=Log[x,1+x]). And, OK, it's not obvious why one would want to do that, and I'm not rushing to make S a built-in function in the Wolfram Language.

But Schönfinkel explains that for him "the practical use of the function S will be to enable us to reduce the number of occurrences of a variable—and to some extent also of a particular function—from several to a single one".

Setting up everything in terms of five basic objects I, C (now K), T, Z and S might already seem impressive and minimalist enough. But Schönfinkel realized that he could go even further:

> Es wird sich für die Durchführung unseres logisch-symbolischen Problems als belangreich erweisen, daß die oben erklärten fünf individuellen Funktionen I, C, T, Z, S des Funktionenkalküls nicht voneinander unabhängig sind, vielmehr zwei von ihnen, nämlich C und S, hinreichen, um die übrigen durch sie zu definieren. Und zwar bestehen hier die folgenden Zusammenhänge:
>
> 1. Es ist gemäß der Erklärung der Funktionen I und C:
> $$Ix = x = Cxy.$$

> Da y willkürlich ist, können wir dafür ein beliebiges Ding oder eine beliebige Funktion einsetzen, also z. B. Cx. Dies gibt:
> $$Ix = (Cx)(Cx).$$
> Nach der Erklärung von S bedeutet dies aber:
> $$SCCx,$$
> so daß wir erhalten:
> $$I = SCC.\ ^3)$$
> Übrigens kommt es in dem Ausdruck SCC auf das letzte Zeichen C gar nicht einmal an. Setzen wir nämlich oben für y nicht Cx, sondern die willkürliche Funktion φx, so ergibt sich entsprechend:
> $$I = SC\varphi,$$
> wo also für φ jede beliebige Funktion eingesetzt werden kann ^{4}).
> 2. Nach der Erklärung von Z ist
> $$Zfgx = f(gx).$$
> Weiter ist vermöge der bereits verwendeten Umformungen:
> $$f(gx) = (Cfx)(gx) = S(Cf)gx = (CSf)(Cf)gx.$$
> Verschmelzung nach f ergibt:
> $$S(CS)Cfgx,$$
> also
> $$Z = S(CS)C.$$
> 3. Ganz entsprechend läßt sich
> $$Tfyx = fxy$$
> weiter umformen in:
> $$fx(Cyx) = (fx)(Cyx) = Sf(Cy)x = (Sf)(Cy)x = Z(Sf)Cyx$$
> $$= ZZSfCyx = (ZZSf)Cyx = (ZZSf)(CCf)yx = S(ZZS)(CC)fyx.$$
> Es gilt somit:
> $$T = S(ZZS)(CC).$$
> Setzt man hier für Z den oben gefundenen Ausdruck ein, so ist damit T ebenfalls auf C und S zurückgeführt.

First, he says that actually $I = SCC$ (or, in modern notation, s[k][k]). In other words, s[k][k][x] for symbolic x is just equal to x (since s[k][k][x] becomes k[x][k[x]] by using the definition of S, and this becomes x by using the definition of C). He notes that this particular reduction was communicated to him by a certain Alfred Boskowitz (who we know to have been a student at the time); he says that Paul Bernays (who was more of a colleague) had "some time before" noted that $I = (SC)(CC)$ (i.e. s[k][k[k]]). Today, of course, we can use a computer to just enumerate all possible combinator expressions of a particular size, and find what the smallest reduction is. But in Schönfinkel's day, it would have been more like solving a puzzle by hand.

Schönfinkel goes on, and proves that Z can also be reduced: $Z = S(CS)C$ (i.e. s[k[s]][k]). And, yes, a very simple Wolfram Language program can verify in a few milliseconds that that is the simplest form.

OK, what about T? Schönfinkel gives 8 steps of reduction to prove that $T = S(ZZS)(CC)$ (i.e. s[s[k[s]][k][s[k[s]][k]][s]][k[k]]). But is this the simplest possible form for T? Well, no. But (with the very straightforward 2-line Wolfram Language program I wrote) it did take my modern computer a number of minutes to determine what the simplest form is.

The answer is that it doesn't have size 12, like Schönfinkel's, but rather size 9. Actually, there are 6 cases of size 9 that all work: s[s[k[s]][s[k[k]][s]]][k[k]] $(S(S(KS)(S(KK)S))(KK)))$ and five others. And, yes, it takes a few steps of reduction to prove that they work (the other size-9 cases $S(SSK(K(SS(KK))))S$, $S(S(K(S(KS)K))S)(KK)$, $S(K(S(S(KS)K)(KK)))S$, $S(K(SS(KK)))(S(KK)S)$, $S(K(S(K(SS(KK)))K))S$ all have more complicated reductions):

```
s[s[k[s]][s[k[k]][s]]][k[k]][f][g][x]
s[k[s]][s[k[k]][s]][f][k[k][f]][g][x]
k[s][f][s[k[k]][s][f]][k[k][f]][g][x]
s[s[k[k]][s][f]][k[k][f]][g][x]
s[k[k]][s][f][g][k[k][f][g]][x]
k[k][f][s[f]][g][k[k][f][g]][x]
k[s[f]][g][k[k][f][g]][x]
s[f][k[k][f][g]][x]
f[x][k[k][f][g][x]]
f[x][k[g][x]]
f[x][g]
```

But, OK, what did Schönfinkel want to do with these objects he'd constructed? As the title of his paper suggests, he wanted to use them as building blocks for mathematical logic. He begins: "Let us now apply our results to a special case, that of the calculus of logic in which the basic elements are individuals and the functions are propositional functions." I consider this sentence significant. Schönfinkel didn't have a way to express it (the concept of universal computation hadn't been invented yet), but he seems to have realized that what he'd done was quite general, and went even beyond being able to represent a particular kind of logic.

Still, he went on to give his example. He'd explained at the beginning of the paper that the quantifiers we now call ∀ and ∃ could both be represented in terms of a kind of "quantified NAND" that he wrote $|^x$:

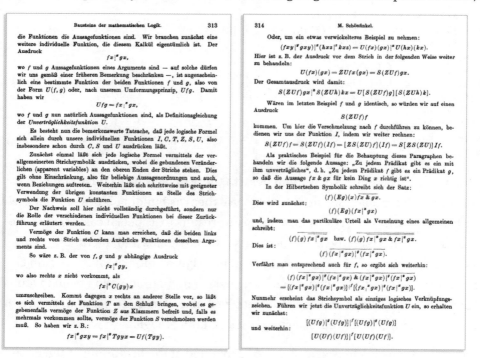

But now he wanted to "combinator-ify" everything. So he introduced a new combinator U, and defined it to represent his "quantified NAND": $Ufg = fx \mid^x gx$ (he called U the "incompatibility function"—an interesting linguistic description of NAND):

"It is a remarkable fact", he says, "that every formula of logic can now be expressed by means... solely of C, S and U." So he's saying that any expression from mathematical logic can be written out as some combinator expression in terms of S, C (now K) and U. He says that when there are quantifiers like "for all x..." it's always possible to use combinators to get rid of the "bound variables" x, etc. He says that he "will not give the complete demonstration here", but rather content himself with an example. (Unfortunately—for reasons of the trajectory of his life that are still quite unclear—he never published his "complete demonstration".)

But, OK, so what had he achieved? He'd basically shown that any expression that might appear in predicate logic (with logical connectives, quantifiers, variables, etc.) could be reduced to an expression purely in terms of the combinators S, C (now K) and U.

Did he need the U? Not really. But he had to have some way to represent the thing with mathematical or logical "meaning" on which his combinators would be acting. Today the obvious thing to do would be to have a representation for true and false. And what's more, to represent these purely in terms of combinators. For example, if we took K to represent true, and SK (s[k]) to represent false, then AND can be represented as SSK (s[s][k]), OR as $S(SS)S(SK)$ (s[s[s]][s][s[k]]) and NAND as $S(S(K(S(SS(K(KK))))))S$ (s[s[k[s[s[s][k[k[k]]]]]]][s]). Schönfinkel got amazingly far in reducing everything to his "building blocks". But, yes, he missed this final step.

But given that he'd managed to reduce everything to S, C and U he figured he should try to go further. So he considered an object J that would be a single building block of S and C: $JJ = S$ and $J(JJ) = C$.

> Weiter als bis zu den Symbolen C, S und U läßt sich, soviel wir sehen, die Zurückführung nicht ohne Zwang treiben.
>   Rein schematisch könnte man freilich C, S und U sogar durch eine einzige Funktion ersetzen, indem man die neue Funktion J einführte durch die Festsetzung:
> $$JC = U, \quad JS = C, \quad Jx = S,$$
> wo x jedes von C und S verschiedene Ding ist. Wir stellen zunächst fest, daß J seinerseits von C und S verschieden ist, da nämlich J nur drei, C ebenso wie S dagegen unendlich viele Funktionswerte annimmt. Wir haben infolgedessen:
> $$JJ = S, \quad J(JJ) = JS = C, \quad J[J(JJ)] = JC = U,$$
> womit die Zurückführung in der Tat geleistet ist. Doch hat diese wegen ihrer augenscheinlichen Willkür wohl kaum sachliche Bedeutung.

With S and K one can just point to any piece of an expression and see if it reduces. With J it's a bit more complicated. In modern Wolfram Language terms one can state the rules as {j[j][x_][y_][z_] → x[z][y[z]], j[j[j]][x_][y_] → x} (where order matters) but to apply these requires pattern matching "clusters of J's" rather than just looking at single S's and K's at a time.

But even though—as Schönfinkel observed—this "final reduction" to J didn't work out, getting everything down to S and K was already amazing. At the beginning of the paper, Schönfinkel had described his objectives. And then he says "It seems to me remarkable in the extreme that the goal we have just set can be realized also; as it happens, it can be done by a reduction to three fundamental signs." (The paper does say three fundamental signs, presumably counting U as well as S and K.)

I'm sure Schönfinkel expected that to reproduce all the richness of mathematical logic he'd need quite an elaborate set of building blocks. And certainly people like Frege, Whitehead and Russell had used what were eventually very complicated setups. Schönfinkel managed to cut through all the complexity to show that simple building blocks were all that were needed. But then he found something else: that actually just two building blocks (S and K) were enough.

In modern terms, we'd say that Schönfinkel managed to construct a system capable of universal computation. And that's amazing in itself. But even more amazing is that he found he could do it with such a simple setup.

I'm sure Schönfinkel was extremely surprised. And here I personally feel a certain commonality with him. Because in my own explorations of the computational universe, what I've found over and over again is that it takes only remarkably simple systems to be capable of highly complex behavior—and of universal computation. And even after exploring the computational universe for four decades, I'm still continually surprised at just how simple the systems can be.

For me, this has turned into a general principle—the Principle of Computational Equivalence—and a whole conceptual framework around it. Schönfinkel didn't have anything like that to think in terms of. But he was in a sense a good enough scientist that he still managed to discover what he discovered—that many decades later we can see fits in as another piece of evidence for the Principle of Computational Equivalence.

Looking at Schönfinkel's paper a century later, it's remarkable not only for what it discovers, but also for the clarity and simplicity with which it is presented. A little of the notation is now dated (and of course the original paper is written in German, which is no longer the kind of leading language of scholarship it once was). But for the most part, the paper still seems perfectly modern. Except, of course, that now it could be couched in terms of symbolic expressions and computation, rather than mathematical logic.

What Is Their Mathematics?

Combinators are hard to understand, and it's not clear how many people understood them when they were first introduced—let alone understood their implications. It's not a good sign that when Schönfinkel's paper appeared in 1924 the person who helped prepare it for final publication (Heinrich Behmann) added his own three paragraphs at the end, that were quite confused. And Schönfinkel's sole other published paper—coauthored with Paul Bernays in 1927—didn't even mention combinators, even though they could have very profitably been used to discuss the subject at hand (decision problems in mathematical logic).

But in 1927 combinators (if not perhaps Schönfinkel's recognition for them) had a remarkable piece of good fortune. Schönfinkel's paper was discovered by a certain Haskell Curry—who would then devote more than 50 years to studying what he named "combinators", and to spreading the word about them.

At some level I think one can view the main thrust of what Curry and his disciples did with combinators as an effort to "mathematicize" them. Schönfinkel had presented combinators in a rather straightforward "structural" way. But what was the mathematical interpretation of what he did, and of how combinators work in general? What mathematical formalism could capture Schönfinkel's structural idea of substitution? Just what, for example, was the true notion of equality for combinators?

In the end, combinators are fundamentally computational constructs, full of all the phenomena of "unbridled computation"—like undecidability and computational irreducibility. And it's inevitable that mathematics as normally conceived can only go so far in "cracking" them.

But back in the 1920s and 1930s the concept and power of computation was not yet understood, and it was assumed that the ideas and tools of mathematics would be the ones to use in analyzing a formal system like combinators. And it wasn't that mathematical methods got absolutely nowhere with combinators.

Unlike cellular automata, or even Turing machines, there's a certain immediate structural complexity to combinators, with their elaborate tree structures, equivalences and so on. And so there was progress to be made—and years of work to be done—in untangling this, without having to face the raw features of full-scale computation, like computational irreducibility.

In the end, combinators are full of computational irreducibility. But they also have layers of computational reducibility, some of which are aligned with the kinds of things mathematics and mathematical logic have been set up to handle. And in this there's a curious resonance with our recent Physics Project.

In our models based on hypergraph rewriting there's also a kind of bedrock of computational irreducibility. But as with combinators, there's a certain immediate structural complexity to what our models do. And there are layers of computational reducibility associated with this. But the remarkable thing with our models is that some of those layers—and the formalisms one can build to understand them—have an immediate interpretation: they are basically the core theories of twentieth-century physics, namely general relativity and quantum mechanics.

Combinators work sufficiently differently that they don't immediately align with that kind of interpretation. But it's still true that one of the important properties discovered in combinators (namely confluence, related to our idea of causal invariance) turns out to be crucial to our models, their correspondence with physics, and in the end our whole ability to perceive regularity in the universe, even in the face of computational irreducibility.

But let's get back to the story of combinators as it played out after Schönfinkel's paper. Schönfinkel had basically set things up in a novel, very direct, structural way. But Curry wanted to connect with more traditional ideas in mathematical logic, and mathematics in general. And after a first paper (published in 1929) which pretty much just recorded his first thoughts, and his efforts to understand what Schönfinkel had done, Curry was by 1930 starting to do things like formulate axioms for combinators, and hoping to prove general theorems about mathematical properties like equality.

Without the understanding of universal computation and their relationship to it, it wasn't clear yet how complicated it might ultimately be to deal with combinators. And Curry pushed forward, publishing more papers and trying to do things like define set theory using his axioms for combinators. But in 1934 disaster struck. It wasn't something about computation or undecidability; instead it was that Stephen Kleene and J. Barkley Rosser showed the axioms Curry had come up with to try and "tighten up Schönfinkel" were just plain inconsistent.

To Kleene and Rosser it provided more evidence of the need for Russell's (originally quite hacky) idea of types—and led them to more complicated axiom systems, and away from combinators. But Curry was undeterred. He revised his axiom system and continued—ultimately for many decades—to see what could be proved about combinators and things like them using mathematical methods.

But already at the beginning of the 1930s there were bigger things afoot around mathematical logic—which would soon intersect with combinators.

Gödel's Theorem and Computability

How should one represent the fundamental constructs of mathematics? Back in the 1920s nobody thought seriously about using combinators. And instead there were basically three "big brands": *Principia Mathematica*, set theory and Hilbert's program. Relations were being found, details were being filled in, and issues were being found. But there was a general sense that progress was being made.

Quite where the boundaries might lie wasn't clear. For example, could one specify a way to "construct any function" from lower-level primitives? The basic idea of recursion was very old (think: Fibonacci). But by the early 1920s there was a fairly well-formalized notion of "primitive recursion" in which functions always found their values from earlier values. But could all "mathematical" functions be constructed this way?

By 1926 it was known that this wouldn't work: the Ackermann function was a reasonable "mathematical" function, but it wasn't primitive recursive. It meant that definitions had to be generalized (e.g. to "general recursive functions" that didn't just look back at earlier values, but could "look forward until..." as well). But there didn't seem to be any fundamental problem with the idea that mathematics could just "mechanistically" be built out forever from appropriate primitives.

But in 1931 came Gödel's theorem. There'd been a long tradition of identifying paradoxes and inconsistencies, and finding ways to patch them by changing axioms. But Gödel's theorem was based on Peano's by-then-standard axioms for arithmetic (branded by Gödel as a fragment of *Principia Mathematica*). And it showed there was a fundamental problem.

In essence, Gödel took the paradoxical statement "this statement is unprovable" and showed that it could be expressed purely as a statement of arithmetic—roughly a statement about the existence of solutions to appropriate integer equations. And basically what Gödel had to do to achieve this was to create a "compiler" capable of compiling things like "this statement is unprovable" into arithmetic.

In his paper one can basically see him building up different capabilities (e.g. representing arbitrary expressions as numbers through Gödel numbering, checking conditions using general recursion, etc.)—eventually getting to a "high enough level" to represent the statement he wanted:

182   Kurt Gödel,

1. $x/y \equiv (Ez) [z \leq x \, \& \, x = y . z]$
x ist teilbar durch y.

2. $\mathrm{Prim}\,(x) \equiv \overline{(Ez)} [z \leq x \, \& \, z \neq 1 \, \& \, z \neq x \, \& \, x/z] \, \& \, x > 1$
x ist Primzahl.

3. $0\, Pr\, x \equiv 0$
$(n+1)\, Pr\, x \equiv \varepsilon y \, [y \leq x \, \& \, \mathrm{Prim}\,(y) \, \& \, x/y \, \& \, y > n\, Pr\, x]$
$n\, Pr\, x$ ist die n-te (der Größe nach) in x enthaltene Primzahl.

4. $0! \equiv 1$
$(n+1)! \equiv (n+1) . n!$

5. $Pr\,(0) \equiv 0$
$Pr\,(n+1) \equiv \varepsilon y \, [y \leq \{Pr\,(n)\}! + 1 \, \& \, \mathrm{Prim}\,(y) \, \& \, y > Pr\,(n)]$
$Pr\,(n)$ ist die n-te Primzahl (der Größe nach).

6. $n\, Gl\, x \equiv \varepsilon y \, [y \leq x \, \& \, x/(n\, Pr\, x)^y \, \& \, \overline{x/(n\, Pr\, x)^{y+1}}]$
$n\, Gl\, x$ ist das n-te Glied der Zahl x zugeordneten Zahlenreihe (für $n > 0$ und n nicht größer als die Länge dieser Reihe).

7. $l(x) \equiv \varepsilon y \, [y \leq x \, \& \, y\, Pr\, x > 0 \, \& \, (y+1)\, Pr\, x = 0]$
$l(x)$ ist die Länge der x zugeordneten Zahlenreihe.

8. $x * y \equiv \varepsilon z \, [z \leq [Pr\,(l(x)+l(y))]^{x+y} \, \&$
$\qquad (n) [n \leq l(x) \rightarrow n\, Gl\, z = n\, Gl\, x] \, \&$
$\qquad (n) [0 < n \leq l(y) \rightarrow (n + l(x))\, Gl\, z = n\, Gl\, y]]$
$x * y$ entspricht der Operation des „Aneinanderfügens" zweier endlicher Zahlenreihen.

9. $R\,(x) \equiv 2^x$
$R\,(x)$ entspricht der nur aus der Zahl x bestehenden Zahlenreihe (für $x > 0$).

10. $E\,(x) \equiv R\,(11) * x * R\,(13)$
$E\,(x)$ entspricht der Operation des „Einklammerns" [11 und 13 sind den Grundzeichen $_n($ und $_n)$ zugeordnet].

11. $n\, \mathrm{Var}\, x \equiv (Ez) [13 < z \leq x \, \& \, \mathrm{Prim}\,(z) \, \& \, x = z^n] \, \& \, n \neq 0$
x ist eine Variable n-ten Typs.

12. $\mathrm{Var}\,(x) \equiv (En) [n \leq x \, \& \, n\, \mathrm{Var}\, x]$
x ist eine Variable.

13. $\mathrm{Neg}\,(x) \equiv R\,(5) * E\,(x)$
$\mathrm{Neg}\,(x)$ ist die Negation von x.

---

^{33}) Das Zeichen $\equiv$ wird im Sinne von „Definitionsgleichheit" verwendet, vertritt also bei Definitionen entweder $=$ oder $\sim$ (im übrigen ist die Symbolik die Hilbertsche).

^{34}) Überall, wo in den folgenden Definitionen eines der Zeichen (x), (Ex), εx auftritt, ist es von einer Abschätzung für x gefolgt. Diese Abschätzung dient lediglich dazu, um die rekursive Natur des definierten Begriffs (vgl. Satz IV) zu sichern. Dagegen würde sich der Umfang der definierten Begriffe durch Weglassung dieser Abschätzung meistens nicht ändern.

^{34a}) Für $0 < n \leq z$, wenn z die Anzahl der verschiedenen in x aufgehenden Primzahlen ist. Man beachte, daß für $n = z+1$ $n\, Pr\, x = 0$ ist!

---

184   Kurt Gödel,

25. $v\, Fr\, n, x \equiv \mathrm{Var}\,(v) \, \& \, \mathrm{Form}\,(x) \, \& \, v = n\, Gl\, x \, \&$
$\qquad n \leq l(x) \, \& \, v\, \overline{\mathrm{Geb}}\, n, x$
Die Variable v ist in x an n-ter Stelle frei.

26. $v\, Fr\, x \equiv (En) [n \leq l(x) \, \& \, v\, Fr\, n, x]$
v kommt in x als freie Variable vor.

27. $Su\, x \binom{n}{y} \equiv \varepsilon z \, [z \leq [Pr\,(l(x) + l(y))]^{x+y} \, \& \, [(Eu, v)\, u, v \leq x \, \&$
$\qquad x = u * R\,(n\, Gl\, x) * v \, \& \, z = u * y * v \, \& \, n = l(u) + 1]]$
$Su\, x \binom{n}{y}$ entsteht aus x, wenn man an Stelle des n-ten Gliedes von $x\, y$ einsetzt (vorausgesetzt, daß $0 < n \leq l(x)$).

28. $0\, St\, v, x \equiv \varepsilon n \, [n \leq l(x) \, \& \, v\, Fr\, n, x$
$\qquad \& \, \overline{(Ep)} [n < p \leq l(x) \, \& \, v\, Fr\, p, x]]$
$(k+1)\, St\, v, x \equiv \varepsilon n \, [n < k\, St\, v, x \, \& \, v\, Fr\, n, x$
$\qquad \& \, \overline{(Ep)} [n < p < k\, St\, v, x \, \& \, v\, Fr\, p, x]]$
$k\, St\, v, x$ ist die $k+1$-te Stelle in x (vom Ende der Formel x an gezählt), an der v in x frei ist (und 0, falls es keine solche Stelle gibt).

29. $A\,(v, x) \equiv \varepsilon n \, [n \leq l(x) \, \& \, n\, St\, v, x = 0]$
$A\,(v, x)$ ist die Anzahl der Stellen, an denen v in x frei ist.

30. $Sb_0 \left(x \begin{smallmatrix} v \\ y \end{smallmatrix}\right) \equiv x$
$Sb_{k+1} \left(x \begin{smallmatrix} v \\ y \end{smallmatrix}\right) \equiv Su \, [Sb_k \left(x \begin{smallmatrix} v \\ y \end{smallmatrix}\right)] \binom{k\, St\, v, x}{y}$

31. $Sb \left(x \begin{smallmatrix} v \\ y \end{smallmatrix}\right) \equiv Sb_{A\,(v, x)} \left(x \begin{smallmatrix} v \\ y \end{smallmatrix}\right)$
$Sb \left(x \begin{smallmatrix} v \\ y \end{smallmatrix}\right)$ ist der oben definierte Begriff Subst $a \binom{v}{b}$.

32. $x\, \mathrm{Imp}\, y \equiv [\mathrm{Neg}\,(x)]\, \mathrm{Dis}\, y$
$x\, \mathrm{Con}\, y \equiv \mathrm{Neg}\,\{[\mathrm{Neg}\,(x)]\, \mathrm{Dis}\, [\mathrm{Neg}\,(y)]\}$
$x\, \mathrm{Aeq}\, y \equiv (x\, \mathrm{Imp}\, y)\, \mathrm{Con}\, (y\, \mathrm{Imp}\, x)$
$v\, \mathrm{Ex}\, y \equiv \mathrm{Neg}\,\{v\, \mathrm{Gen}\, [\mathrm{Neg}\,(y)]\}$

33. $n\, Th\, x \equiv \varepsilon y \, [y \leq x^{x^{n}} \, \& \, (k) [k \leq l(x) \rightarrow$
$\qquad (k\, Gl\, x \leq 13 \, \& \, k\, Gl\, y = k\, Gl\, x) \, \vee$
$\qquad (k\, Gl\, x > 13 \, \& \, k\, Gl\, y = k\, Gl\, x . [1\, Pr\,(k\, Gl\, x)^n])]]$
$n\, Th\, x$ ist die n-te Typenerhöhung von x (falls x und $n\, Th\, x$ Formeln sind).

Den Axiomen I, 1 bis 3 entsprechen drei bestimmte Zahlen, die wir mit z_1, z_2, z_3 bezeichnen, und wir definieren:

34. $Z\text{-}Ax\,(x) \equiv (x = z_1 \, \vee \, x = z_2 \, \vee \, x = z_3)$

---

^{35}) Falls v keine Variable oder x keine Formel ist, ist $Sb \left(x \begin{smallmatrix} v \\ y \end{smallmatrix}\right) = x$.

^{37}) Statt $Sb \, [Sb \left(x \begin{smallmatrix} v \\ y \end{smallmatrix}\right) \begin{smallmatrix} w \\ z \end{smallmatrix}]$ schreiben wir: $Sb \left(x \begin{smallmatrix} v & w \\ y & z \end{smallmatrix}\right)$ (analog für mehr als zwei Variable).

What did Gödel's theorem mean? For the foundations of mathematics it meant that the idea of mechanically proving "all true theorems of mathematics" wasn't going to work. Because it showed that there was at least one statement that by its own admission couldn't be proved, but was still a "statement about arithmetic", in the sense that it could be "compiled into arithmetic".

That was a big deal for the foundations of mathematics. But actually there was something much more significant about Gödel's theorem, even though it wasn't recognized at the time. Gödel had used the primitives of number theory and logic to build what amounted to a computational system—in which one could take things like "this statement is unprovable", and "run them in arithmetic".

What Gödel had, though, wasn't exactly a streamlined general system (after all, it only really needed to handle one statement). But the immediate question then was: if there's a problem with this statement in arithmetic, what about Hilbert's general "decision problem" (*Entscheidungsproblem*) for any axiom system?

To discuss the "general decision problem", though, one needed some kind of general notion of how one could decide things. What ultimate primitives should one use? Schönfinkel (with Paul Bernays)—in his sole other published paper—wrote about a restricted case of the decision problem in 1927, but doesn't seem to have had the idea of using combinators to study it.

By 1934 Gödel was talking about general recursiveness (i.e. definability through general recursion). And Alonzo Church and Stephen Kleene were introducing λ definability. Then in 1936 Alan Turing introduced Turing machines. All these approaches involved setting up certain primitives, then showing that a large class of things could be "compiled" to those primitives. And that—in effect by thinking about having it compile itself—Hilbert's *Entscheidungsproblem* couldn't be solved.

Perhaps no single result along these lines would have been so significant. But it was soon established that all three kinds of systems were exactly equivalent: the set of computations they could represent were the same, as established by showing that one system could emulate another. And from that discovery eventually emerged the modern notion of universal computation—and all its implications for technology and science.

In the early days, though, there was actually a fourth equivalent kind of system—based on string rewriting—that had been invented by Emil Post in 1920–1. Oh, and then there were combinators.

Lambda Calculus

What was the right "language" to use for setting up mathematical logic? There'd been gradual improvement since the complexities of *Principia Mathematica*. But around 1930 Alonzo Church wanted a new and cleaner setup. And he needed to have a way (as Frege and *Principia Mathematica* had done before him) to represent "pure functions". And that's how he came to invent λ.

Today in the Wolfram Language we have Function[x, f[x]] or x↦f[x] (or various shorthands). Church originally had $\lambda x[\mathbf{M}]$:

But what's perhaps most notable is that on the very first page he defines λ, he's referencing Schönfinkel's combinator paper. (Well, specifically, he's referencing it because he wants to use the device Schönfinkel invented that we now call currying—f[x][y] in place of f[x,y]—though ironically he doesn't mention Curry.) In his 1932 paper (apparently based on work in 1928–9) λ is almost a sideshow—the main event being the introduction of 37 formal postulates for mathematical logic:

356   A. CHURCH.

III. *If I is true, if M and N are well-formed, if the variable x occurs in M,*
and if the bound variables in M are distinct both from x and from
the free variable in N, then K, the result of substituting {λx.M} (N)
for a particular occurrence of Sᴹₓ |M| in I, is also true.
IV. *If {F} (A) is true and F and A are well-formed, then Σ(F) is true.*
V. *If Π(F, G) are true, and F, G, and A are well-formed,*
then {G} (A) is true.
And our formal postulates are the thirty-seven following:

1. $\Sigma(\varphi) \supset_\varphi \Pi(\psi, \varphi)$.
2. ${}'x.\varphi(x) \supset_\varphi . \Pi(\psi, \psi) \supset_\psi \psi(x)$.
3. $\Sigma(\sigma) \supset_\varphi . [\sigma(x) \supset_x \psi(x)] \supset_\varphi . \Pi(\psi, \varphi) \supset_\psi . \sigma(x) \supset_x \psi(x)$.
4. $\Sigma(x) \supset_\varphi . \Sigma y [\varphi(x) \supset_x \psi(x, y)] \supset_y . [\varphi(x) \supset_x \Pi(\psi(x), \psi(x))] \supset_\psi .$
   $[\varphi(x) \supset_x \psi(x, y)] \supset_y . \varphi(x) \supset_x \psi(x, y)$.
5. $\Sigma(\varphi) \supset_\varphi . \Pi(\varphi, \psi) \supset_\psi . \varphi(f(x)) \supset_{f_x} \psi(f(x))$.
6. ${}'x.\varphi(x) \supset_\varphi . \Pi(\psi, \varphi(x)) \supset_\psi \psi(x, x)$.
7. $\varphi(x, f(x)) \supset_{\varphi f x} . \Pi(\psi(x), \psi(x)) \supset_\psi \psi(x, f(x))$.
8. $\Sigma(x) \supset_\varphi . \Sigma y [\varphi(x) \supset_x \varphi(x, y)] \supset_\varphi . [\varphi(x) \supset_x \Pi(\psi(x), \psi)] \supset_\psi .$
   $[\varphi(x) \supset_x \varphi(x, y)] \supset_y \psi(y)$.
9. ${}'x.\varphi(x) \supset_\varphi \Sigma(\varphi)$.
10. $\Sigma x \varphi(f(x)) \supset_{f\varphi} \Sigma(\varphi)$.
11. $\varphi(x, x) \supset_{\varphi x} \Sigma(\varphi(x))$.
12. $\Sigma(\varphi) \supset_\varphi \Sigma x \varphi(x)$.
13. $\Sigma(\varphi) \supset_\varphi . [\varphi(x) \supset_x \psi(x)] \supset_\psi \Pi(\varphi, \psi)$.
14. $p \supset_p . q \supset_q pq$.
15. $pq \supset_{pq} p$.
16. $pq \supset_{pq} q$.
17. $\Sigma x \Sigma \theta [\varphi(x) . \sim \theta(x) . \Pi(\psi, \theta)] \supset_{\varphi\psi} \sim \Pi(\varphi, \psi)$.
18. $\sim \Pi(\varphi, \psi) \supset_{\varphi\psi} \Sigma x \Sigma \theta . \varphi(x) . \sim \theta(x) . \Pi(\psi, \theta)$.
19. $\Sigma x \Sigma \theta [\sim \varphi(u, x) . \sim \theta(u) . \Sigma(\varphi(y)) \supset_y \theta(y)] \supset_{\varphi u} \sim \Sigma(\varphi(u))$.
20. $\sim \Sigma(\varphi) \supset_\varphi \Sigma x . \sim \varphi(x)$.
21. $p \supset_p . \sim q \supset_q \sim . pq$.
22. $\sim p \supset_p . q \supset_q \sim . pq$.
23. $\sim p \supset_p . \sim q \supset_q \sim . pq$.
24. $p \supset_p . |\sim . pq| \supset_q \sim q$.

By the next year J. Barkley Rosser is trying to retool Curry's "combinatory logic" with
combinators of his own—and showing how they correspond to lambda expressions:

We shall use Church's method for denoting definitions (see Church 1932, p.
355) and shall list the following, giving on the right the equivalent in Church's
notation:

$T \to JII$	$\lambda x f \cdot f(x)$	(See footnote 4)
$C \to JT(JT)(JT)$	$\lambda f x y \cdot f(y, z)$	
$B \to C(JIC)(JI)$	$\lambda f g x \cdot f(g(x))$	
$W \to C(C(BC(C(BJT)T))T)$	$\lambda f x \cdot f(x, x)$	
$1 \to BI$	$\lambda f x \cdot f(x)$	
$\mathbf{p} \times \mathbf{q} \to B\mathbf{pq}$	$\lambda x \cdot \mathbf{p}(\mathbf{q}(x))$	

⁴ $\lambda x f \cdot f(x)$ is an abbreviation of $\lambda x \lambda f \cdot f(x)$.

Then in 1935 lambda calculus has its big "coming out" in Church's "An Unsolv-
able Problem of Elementary Number Theory", in which he introduces the idea
that any "effectively calculable" function should be "λ definable", then defines
integers in terms of λ's ("Church numerals")

We introduce at once the following infinite list of abbreviations,

$$1 \to \lambda ab \cdot a(b),$$
$$2 \to \lambda ab \cdot a(a(b)),$$
$$3 \to \lambda ab \cdot a(a(a(b))),$$

and then shows that the problem of determining equivalence for λ expressions is undecidable.

Very soon thereafter Turing publishes his "On Computable Numbers, with an
Application to the *Entscheidungsproblem*" in which he introduces his much more
manifestly mechanistic Turing machine model of computation. In the main part of

the paper there are no lambdas—or combinators—to be seen. But by late 1936 Turing had gone to Princeton to be a student with Church—and added a note showing the correspondence between his Turing machines and Church's lambda calculus.

By the next year, when Turing is writing his rather abstruse "Systems of Logic Based on Ordinals" he's using lambda calculus all over the place. Early in the document he writes $I \rightarrow \lambda x[x]$, and soon he's mixing lambdas and combinators with wild abandon—and in fact he'd already published a one-page paper which introduced the fixed-point combinator Θ (and, yes, the K in the title refers to Schönfinkel's K combinator):

When Church summarized the state of lambda calculus in 1941 in his "The Calculi of Lambda-Conversion" he again made extensive use of combinators. Schönfinkel's K is prominent. But Schönfinkel's S is nowhere to be seen—and in fact Church has his own S combinator $S[n][f][x] \rightarrow f[n[f][x]]$ which implements successors in Church's numeral system. And he also has a few other "basic combinators" that he routinely uses.

In the end, combinators and lambda calculus are completely equivalent, and it's quite easy to convert between them—but there's a curious tradeoff. In lambda calculus one names variables, which is good for human readability, but can lead to problems at a formal level. In combinators, things are formally much cleaner, but the expressions one gets can be completely incomprehensible to humans.

The point is that in a lambda expression like $\lambda x\, \lambda y\, x[y]$ one's naming the variables (here x and y), but really these names are just placeholders: what they are doesn't matter; they're just showing where different arguments go. And in a simple case like this, everything is fine. But what happens if one substitutes for y another lambda expression, say $\lambda x\, f[x]$? What is that x? Is it the same x as the one outside, or something different? In practice, there are all sorts of renaming schemes that can be used, but they tend to be quite hacky, and things can quickly get tangled up. And if one wants to make formal proofs about lambda calculus, this can potentially be a big problem, and indeed at the beginning it wasn't clear it wouldn't derail the whole idea of lambda calculus.

And that's part of why the correspondence between lambda calculus and combinators was important. With combinators there are no variables, and so no variable names to get tangled up. So if one can show that something can be converted to combinators—even if one never looks at the potentially very long and ugly combinator expression that's generated—one knows one's safe from issues about variable names.

There are still plenty of other complicated issues, though. Prominent among them are questions about when combinator expressions can be considered equal. Let's say you have a combinator expression, like s[s[s[s][k]]][k]. Well, you can repeatedly apply the rules for combinators to transform and reduce it. And it'll often end up at a fixed point, where no rules apply anymore. But a basic question is whether it matters in which order the rules are applied. And in 1936 Church and Rosser proved it doesn't.

Actually, what they specifically proved was the analogous result for lambda calculus. They drew a picture to indicate different possible orders in which lambdas could be reduced out, and showed it didn't matter which path one takes:

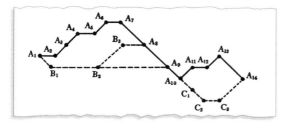

This all might seem like a detail. But it turns out that generalizations of their result apply to all sorts of systems. In doing computations (or automatically proving theorems) it's all about "it doesn't matter what path you take; you'll always get the same result". And that's important. But recently there's been another important application that's shown up. It turns out that a generalization of the "Church–Rosser property" is what we call causal invariance in our Physics Project.

And it's causal invariance that leads in our models to relativistic invariance, general covariance, objective reality in quantum mechanics, and other central features of physics.

Practical Computation

In retrospect, one of the great achievements of the 1930s was the inception of what ended up being the idea of universal computation. But at the time what was done was couched in terms of mathematical logic and it was far from obvious that any of the theoretical structures being built would have any real application beyond thinking about the foundations of mathematics. But even as people like Hilbert were talking in theoretical terms about the mechanization of mathematics, more and more there were actual machines being built for doing mathematical calculations.

We know that even in antiquity (at least one) simple gear-based mechanical calculational devices existed. In the mid-1600s arithmetic calculators started being constructed, and by the late 1800s they were in widespread use. At first they were mechanical, but by the 1930s most were electromechanical, and there started to be systems where units for carrying out different arithmetic operations could be chained together. And by the end of the 1940s fairly elaborate such systems based on electronics were being built.

Already in the 1830s Charles Babbage had imagined an "analytical engine" which could do different operations depending on a "program" specified by punch cards—and Ada Lovelace had realized that such a machine had broad "computational" potential. But by the 1930s a century had passed and nothing like this was connected to the theoretical developments that were going on—and the actual engineering of computational systems was done without any particular overarching theoretical framework.

Still, as electronic devices got more complicated and scientific interest in psychology intensified, something else happened: there started to be the idea (sometimes associated with the name cybernetics) that somehow electronics might reproduce how things like brains work. In the mid-1930s Claude Shannon had shown that Boolean algebra could represent how switching circuits work, and in 1943 Warren McCulloch and Walter Pitts proposed a model of idealized neural networks formulated in something close to mathematical logic terms.

Meanwhile by the mid-1940s John von Neumann—who had worked extensively on mathematical logic—had started suggesting math-like specifications for practical electronic computers, including the way their programs might be stored electronically. At first he made lots of brain-like references to "organs" and "inhibitory connections", and essentially no mention of ideas from mathematical logic. But by the end of the 1940s von Neumann was talking at least conceptually about connections to Gödel's theorem and Turing machines, Alan Turing had become involved with actual electronic computers, and there was the beginning of widespread understanding of the notion of general-purpose computers and universal computation.

In the 1950s there was an explosion of interest in what would now be called the theory of computation—and great optimism about its relevance to artificial intelligence. There was all sorts of "interdisciplinary work" on fairly "concrete" models of computation, like finite automata, Turing machines, cellular automata and idealized neural networks. More "abstract" approaches, like recursive functions, lambda calculus—and combinators—remained, however, pretty much restricted to researchers in mathematical logic.

When early programming languages started to appear in the latter part of the 1950s, thinking about practical computers began to become a bit more abstract. It was understood that the grammars of languages could be specified recursively— and actual recursion (of functions being able to call themselves) just snuck into the specification of ALGOL 60. But what about the structures on which programs operated? Most of the concentration was on arrays (sometimes rather elegantly, as in APL) and, occasionally, character strings.

But a notable exception was LISP, described in John McCarthy's 1960 paper "Recursive Functions of Symbolic Expressions and Their Computation by Machine, Part I" (part 2 was not written). There was lots of optimism about AI at the time, and the idea was to create a language to "implement AI"—and do things like "mechanical theorem proving". A key idea—that McCarthy described as being based on "recursive function formalism"—was to have tree-structured symbolic expressions ("S expressions"). (In the original paper, what's now Wolfram Language–style f[g[x]] "M expression" notation, complete with square brackets, was used as part of the specification, but the quintessential-LISP-like $(f(g\,x))$ notation won out when LISP was actually implemented.)

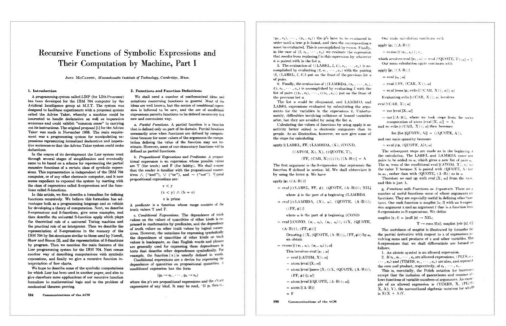

An issue in LISP was how to take "expressions" (which were viewed as representing things) and turn them into functions (which do things). And the basic plan was to use Church's idea of λ notation. But when it came time to implement this, there was, of course, trouble with name collisions, which ended up getting handled in quite hacky ways. So did McCarthy know about combinators? The answer is yes, as his 1960 paper shows:

> Difficulties arise in combining functions described by λ-expressions, or by any other notation involving variables, because different bound variables may be represented by the same symbol. This is called collision of bound variables. There is a notation involving operators that are called combinators for combining functions without the use of variables. Unfortunately, the combinatory expressions for interesting combinations of functions tend to be lengthy and unreadable.

I actually didn't know until just now that McCarthy had ever even considered combinators, and in the years I knew him I don't think I ever personally talked to him about them. But it seems that for McCarthy—as for Church—combinators were a kind of "comforting backstop" that ensured that it was OK to use lambdas, and that if things went too badly wrong with variable naming, there was at least in principle always a way to untangle everything.

In the practical development of computers and computer languages, even lambdas—let alone combinators—weren't really much heard from again (except in a small AI circle) until the 1980s. And even then it didn't help that in an effort variously to stay close to hardware and to structure programs there tended to be a desire to give everything a "data type"—which was at odds with the "consume any expression" approach of standard combinators and lambdas. But beginning in the 1980s—particularly with the progressive rise of functional programming—lambdas, at least, have steadily gained in visibility and practical application.

What of combinators? Occasionally as a proof of principle there'll be a hardware system developed that natively implements Schönfinkel's combinators. Or—particularly in modern times—there'll be an esoteric language that uses combinators in some kind of purposeful effort at obfuscation. Still, a remarkable cross-section of notable people concerned with the foundations of computing have—at one time or another—taught about combinators or written a paper about them. And in recent years the term "combinator" has become more popular as a way to describe a "purely applicative" function.

But by and large the important ideas that first arose with combinators ended up being absorbed into practical computing by quite circuitous routes, without direct reference to their origins, or to the specific structure of combinators.

Combinators in Culture

For 100 years combinators have mostly been an obscure academic topic, studied particularly in connection with lambda calculus, at borders between theoretical computer science, mathematical logic and to some extent mathematical formalisms like category theory. Much of the work that's been done can be traced in one way or another to the influence of Haskell Curry or Alonzo Church—particularly through their students, grandstudents, great-grandstudents, etc. Partly in the early years, most of the work was centered in the US, but by the 1960s there was a strong migration to Europe and especially the Netherlands.

But even with all their abstractness and obscurity, on a few rare occasions combinators have broken into something closer to the mainstream. One such time was with the popular logic-puzzle book *To Mock a Mockingbird,* published in 1985 by Raymond Smullyan—a former student of Alonzo Church's. It begins: "A certain enchanted forest is inhabited by talking birds" and goes on to tell a story that's basically about combinators "dressed up" as birds calling each other (*S* is the "starling", *K* the "kestrel")—with a convenient "bird who's who" at the end. The book is dedicated "To the memory of Haskell Curry—an early pioneer in combinatory logic and an avid bird-watcher".

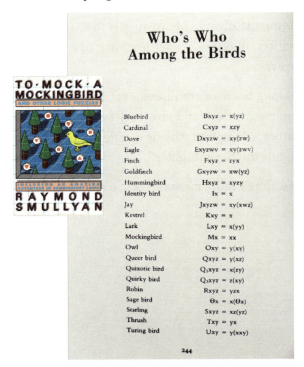

Who's Who Among the Birds

Bluebird	$Bxyz = x(yz)$
Cardinal	$Cxyz = xzy$
Dove	$Dxyzw = xy(zw)$
Eagle	$Exyzwv = xy(zwv)$
Finch	$Fxyz = zyx$
Goldfinch	$Gxyzw = xw(yz)$
Hummingbird	$Hxyz = xyzy$
Identity bird	$Ix = x$
Jay	$Jxyzw = xy(xwz)$
Kestrel	$Kxy = x$
Lark	$Lxy = x(yy)$
Mockingbird	$Mx = xx$
Owl	$Oxy = y(xy)$
Queer bird	$Qxyz = y(xz)$
Quixotic bird	$Q_1xyz = x(zy)$
Quirky bird	$Q_3xyz = z(xy)$
Robin	$Rxyz = yzx$
Sage bird	$\Theta x = x(\Theta x)$
Starling	$Sxyz = xz(yz)$
Thrush	$Txy = yx$
Turing bird	$Uxy = y(xxy)$

244

And then there's Y Combinator. The original *Y* combinator arose out of work that Curry did in the 1930s on the consistency of axiom systems for combinators, and it appeared explicitly in his 1958 classic book:

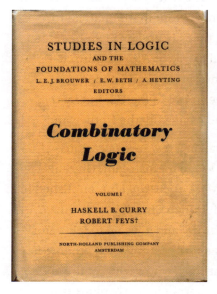

He called it the "paradoxical combinator" because it was recursively defined in a kind of self-referential way analogous to various paradoxes. Its explicit form is $SSK(S(K(SS(S(SSK))))K)$ and its most immediately notable feature is that under Schönfinkel's combinator transformation rules it never settles down to a particular "value" but just keeps growing forever.

Well, in 2005 Paul Graham—who had long been an enthusiast of functional programming and LISP—decided to name his new (and now very famous) startup accelerator "Y Combinator". I remember asking him why he'd called it that. "Because," he said, "nobody understands the *Y* combinator".

Looking in my own archives from that time I find an email I sent a combinator enthusiast who was working with me:

> **Subject: combinators**
> **Date:** Mon, 09 May 2005 00:24:42 -0400
> **From:** Stephen Wolfram
> **To:** Matthew Szudzik
>
>
> From my friend Paul Graham ....
>
> A new application of combinators:
>
> http://www.ycombinator.com/

Followed by, basically, "Yes our theorem prover can prove the basic property of the *Y* combinator" (V6 sounds so ancient; we're now just about to release V12.2):

```
(In a modern V6):

In[29]:=
FullSimplify[Exists[{Y},Y\[Equal]apply[combinator,Y]],
ForAll[{x,y},apply[apply[I,x],y]\[Equal]apply[x,apply[y,y]]]]

Out[29]=
True
```

I had another unexpected encounter with combinators last year. I had been given a book that was once owned by Alan Turing, and in it I found a piece of paper— that I recognized as being covered with none other than lambdas and combinators (but that's not *the Y* combinator):

It took quite a bit of sleuthing (that I wrote extensively about)—but I eventually discovered that the piece of paper was written by Turing's student Robin Gandy. But I never figured out why he was doing combinators....

Designing Symbolic Language

I think I first found out about combinators around 1979 by seeing Schönfinkel's original paper in a book called *From Frege to Gödel: A Source Book in Mathematical Logic* (by a certain Jean van Heijenoort). How Schönfinkel's paper ended up being in that book is an interesting question, which I'll write about elsewhere. The spine of my copy of the book has long been broken at the location of Schönfinkel's paper, and at different times I've come back to the paper, always thinking there was more to understand about it.

But why was I even studying things like this back in 1979? I guess in retrospect I can say I was engaged in an activity that goes back to Frege or even Leibniz: I was trying to find a fundamental framework for representing mathematics and beyond. But my goal wasn't a philosophical one; it was a very practical one: I was trying to build a computer language that could do general computations in mathematics and beyond.

My immediate applications were in physics, and it was from physics that my main methodological experience came. And the result was that—like trying to understand the world in terms of elementary particles—I wanted to understand computation in terms of its most fundamental elements. But I also had lots of practical experience in using computers to do mathematical computation. And I soon developed a theory about how I thought computation could fundamentally be done.

It started from the practical issue of transformations on algebraic expressions (turn $\sin(2x)$ into $2 \sin(x) \cos(x)$, etc.). But it soon became a general idea: compute by doing transformations on symbolic expressions. Was this going to work? I wanted to understand as fundamentally as possible what computation really was— and from that I was led to its history in mathematical logic. Much of what I saw in books and papers about mathematical logic I found abstruse and steeped in sometimes horrendous notational complexity. But what were these people really doing? It made it much easier that I had a definite theory, against which I could essentially do reductionist science. That stuff in *Principia Mathematica*? Those ideas about rewriting systems? Yup, I could see how to represent them as rules for transformations on symbolic expressions.

And so it was that I came to design SMP: "A Symbolic Manipulation Program"—all based on transformation rules for symbolic expressions. It was easy to represent mathematical relations ($x is a pattern variable that would now in the Wolfram Language be x_ on the left-hand side only):

```
STr[3,2,1]:    Asin[$x]+Asin[$y] -> Asin[$x Sqrt[1-$y^2]+$y Sqrt[1-$x^2]]

STr[3,2,2]:    Asin[$x]-Asin[$y] -> Asin[$x Sqrt[1-$y^2]-$y Sqrt[1-$x^2]]

STr[3,2,3]:    Acos[$x]+Acos[$y] -> Acos[$x $y+Sqrt[(1-$x^2) (1-$y^2)]]

STr[3,2,4]:    Acos[$x]-Acos[$y] -> Acos[$x $y-Sqrt[(1-$x^2) (1-$y^2)]]
```

Or basic logic:

```
/* Idempotent laws */
$p | $p : $p
$p & $p : $p

/* Commutative and associative laws built in */

/* Distributive laws */
$p | ($$q & $r) : ($p | $$q) & ($p | $r)

/* Identity laws built in */

/* Complement laws */
~~$p : $p
$p | ~$p : 1
$p & ~$p : 0

/* DeMorgan's laws */
~($$p | $q) : (~$$p) & (~$q)
~($$p & $q) : (~$$p) | (~$q)

/* Reflexive law */
$p => $p : 1

/* Antisymmetric law */
($p => $q) & ($q => $p) : $p=$q

/* Transitive law */
($p => $q) & ($q => $r) : $p=>$r
```

Or, for that matter, predicate logic of the kind Schönfinkel wanted to capture:

```
Quant[$s] : $s
Quant[$$q,Quant[$$r]] : Quant[$$q,$$r]

/* DeMorgan's law */
~Quant[All[$x],$$q,$s] : Quant[Some[$x],~Quant[$$q,$s]]
~Quant[Some[$x],$$q,$s] : Quant[All[$x],~Quant[$$q,$s]]
```

And, yes, it could emulate a Turing machine (note the tape-as-transformation-rules representation that appears at the end):

```
#I[2]::  Tape[$i_$i<$]:1
#O[2]:   1
#I[3]::  Spec[1,0]:{Right,1}
#O[3]:   {Right,1}
#I[4]::  Spec[1,1]:{0,2}
#O[4]:   {0,2}
#I[5]::  Spec[2,0]:{Right,2}
#O[5]:   {Right,2}
#I[6]::  Spec[2,1]:{Right,1}
#O[6]:   {Right,1}
#I[7]::  Start[1,1]
#O[7]:   {S,1}
#I[8]::  Tape
#O[8]:   {[3]: 0, [1]: 0, ($i_ (5 > $i)): 1}
```

But the most important thing I realized is that it really worked to represent basically anything in terms of symbolic expressions, and transformation rules on them. Yes, it was quite often useful to think of "applying functions to things" (and SMP had its version of lambda, for example), but it was much more powerful to think about symbolic expressions as just "being there" ("*x* doesn't have to have a value")—like things in the world—with the language being able to define how things should transform.

In retrospect this all seems awfully like the core idea of combinators, but with one important exception: that instead of everything being built from "purely structural elements" with names like *S* and *K*, there was a whole collection of "primitive objects" that were intended to have direct understandable meanings (like Plus, Times, etc.). And indeed I saw a large part of my task in language design as being to think about computations one might want to do, and then try to "drill down" to find the "elementary particles"—or primitive objects—from which these computations might be built up.

Over time I've come to realize that doing this is less about what one can in principle use to construct computations, and more about making a bridge to the way humans think about things. It's crucial that there's an underlying structure—symbolic expressions—that can represent anything. But increasingly I've come to realize that what we need from a computational language is to have a way to encapsulate in precise computational form the kinds of things we humans think about—in a way that we humans can understand. And a crucial part of being able to do that is to leverage what has ultimately been at the core of making our whole intellectual development as a species possible: the idea of human language.

Human language has given us a way to talk symbolically about the world: to give symbolic names to things, and then to build things up using these. In designing a computational language the goal is to leverage this: to use what humans already know and understand, but be able to represent it in a precise computational way that is amenable to actual computation that can be done automatically by computer.

It's probably no coincidence that the tree structure of symbolic expressions that I have found to be such a successful foundation for computational language is a bit like an idealized version of the kind of tree structure (think parse trees or sentence diagramming) that one can view human language as following. There are other ways to set up universal computation, but this is the one that seems to fit most directly with our way of thinking about things.

And, yes, in the end all those symbolic expressions could be constructed like combinators from objects—like S and K—with no direct human meaning. But that would be like having a world without nouns—a world where there's no name for anything—and the representation of everything has to be built from scratch. But the crucial idea that's central to human language—and now to computational language—is to be able to have layers of abstraction, where one can name things and then refer to them just by name without having to think about how they're built up "inside".

In some sense one can see the goal of people like Frege—and Schönfinkel—as being to "reduce out" what exists in mathematics (or the world) and turn it into something like "pure logic". And the structural part of that is exactly what makes computational language possible. But in my conception of computational language the whole idea is to have content that relates to the world and the way we humans think about it.

And over the decades I've continually been amazed at just how strong and successful the idea of representing things in terms of symbolic expressions and transformations on them is. Underneath everything that's going on in the Wolfram Language—and in all the many systems that now use it—it's all ultimately just symbolic expressions being transformed according to particular rules, and reaching fixed points that represent results of computations, just like in those examples in Schönfinkel's original paper.

One important feature of Schönfinkel's setup is the idea that one doesn't just have "functions" like $f[x]$, or even just nested functions, like $f[g[x]]$. Instead one can have constructs where instead of the "name of a function" (like f) one can have a whole complex symbolic structure. And while this was certainly possible in SMP, not too much was built around it. But when I came to start designing what's now the Wolfram Language in 1986, I made sure that the "head" (as I called it) of an expression could itself be an arbitrary expression.

And when Mathematica was first launched in 1988 I was charmed to see more than one person from mathematical logic immediately think of implementing combinators. Make the definitions:

In[ ]:= s[x_][y_][z_] := x[z][y[z]]

In[ ]:= k[x_][y_] := x

Then combinators "just work" (at least if they reach a fixed point):

In[ ]:= s[s[k[s]][s[k[k]][s[k[s]][k]]]][s[k[s[s[k][k]]]][k]][a][b][c]

Out[ ]:= a[b[a][c]]

But what about the idea of "composite symbolic heads"? Already in SMP I'd used them to do simple things like represent derivatives (and in Wolfram Language f'[x] is Derivative[1][f][x]). But something that's been interesting to me to see is that as the decades have gone by, more and more gets done with "composite heads". Sometimes one thinks of them as some kind of nesting of operations, or nesting of modifiers to a symbolic object. But increasingly they end up being a way to represent "higher-order constructs"—in effect things that produce things that produce things etc. that eventually give a concrete object one wants.

I don't think most of us humans are particularly good at following this kind of chain of abstraction, at least without some kind of "guide rails". And it's been interesting for me to see over the years how we've been able to progressively build up guide rails for longer and longer chains of abstraction. First there were things like Function, Apply, Map. Then Nest, Fold, FixedPoint, MapThread. But only quite recently NestGraph, FoldPair, SubsetMap, etc. Even from the beginning there were direct "head manipulation" functions like Operate and Through. But unlike more "array-like" operations for list manipulation they've been slow to catch on.

In a sense combinators are an ultimate story of "symbolic head manipulation": everything can get applied to everything before it's applied to anything. And, yes, it's very hard to keep track of what's going on—which is why "named guide rails" are so important, and also why they're challenging to devise. But it seems as if, as we progressively evolve our understanding, we're slowly able to get a little further, in effect building towards the kind of structure and power that combinators—in their very non-human-relatable way—first showed us was possible a century ago.

Combinators in the Computational Universe

Combinators were invented for a definite purpose: to provide building blocks, as Schönfinkel put it, for logic. It was the same kind of thing with other models of what we now know of as computation. All of them were "constructed for a purpose". But in the end computation—and programs—are abstract things, that can in principle be studied without reference to any particular purpose. One might have some particular reason to be looking at how fast programs of some kind can run, or what can be proved about them. But what about the analog of pure natural science: of studying what programs just "naturally do"?

At the beginning of the 1980s I got very interested in what one can think of as the "natural science of programs". My interest originally arose out of a question about ordinary natural science. One of the very noticeable features of the natural world is how much in it seems to us highly complex. But where does this complexity really come from? Through what kind of mechanism does nature produce it? I quickly realized that in trying to address that question, I needed as general a foundation for making models of things as possible. And for that I turned to programs, and began to study just what "programs in the wild" might do.

Ever since the time of Galileo and Newton mathematical equations had been the main way that people ultimately imagined making models of nature. And on the face of it—with their real numbers and continuous character—these seemed quite different from the usual setup for computation, with its discrete elements and discrete choices. But perhaps in part through my own experience in doing mathematics symbolically on computers, I didn't see a real conflict, and I began to think of programs as a kind of generalization of the traditional approach to modeling in science.

But what kind of programs might nature use? I decided to just start exploring all the possibilities: the whole "computational universe" of programs—starting with the simplest. I came up with a particularly simple setup involving a row of cells with values 0 or 1 updated in parallel based on the values of their neighbors. I soon learned that systems like this had actually been studied under the name "cellular automata" in the 1950s (particularly in 2D) as potential models of computation, though had fallen out of favor mainly through not having seemed very "human programmable".

My initial assumption was that with simple programs I'd only see simple behavior. But with my cellular automata it was very easy to do actual computer experiments, and to visualize the results. And though in many cases what I saw was simple behavior, I also saw something very surprising: that in some cases—even though the rules were very simple—the behavior that was generated could be immensely complex:

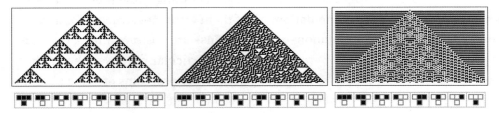

It took me years to come to terms with this phenomenon, and it's gradually informed the way I think about science, computation and many other things. At first I studied it almost exclusively in cellular automata. I made connections to actual systems in nature that cellular automata could model. I tried to understand what existing mathematical and other methods could say about what I'd seen. And slowly I began to formulate general ideas to explain what was going on—like computational irreducibility and the Principle of Computational Equivalence.

But at the beginning of the 1990s—now armed with what would become the Wolfram Language—I decided I should try to see just how the phenomenon I had found in cellular automata would play it in other kinds of computational systems. And my archives record that on April 4, 1992, I started looking at combinators.

I seem to have come back to them several times, but in a notebook from July 10, 1994 (which, yes, still runs just fine), there it is:

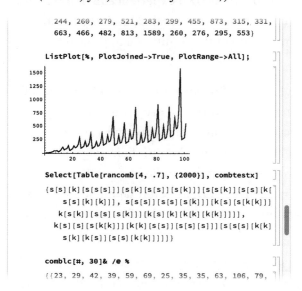

A randomly chosen combinator made of Schönfinkel's *S*'s and *K*'s starting to show complex behavior. I seem to have a lot of notebooks that start with the simple combinator definitions—and then start exploring:

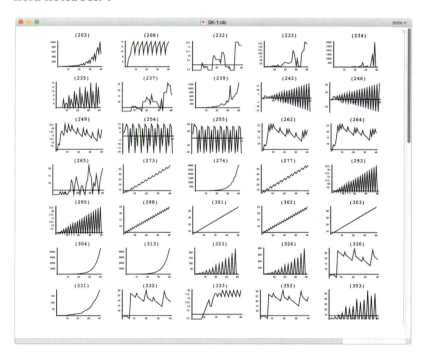

There are what seem like they could be pages from a "computational naturalist's field notebook":

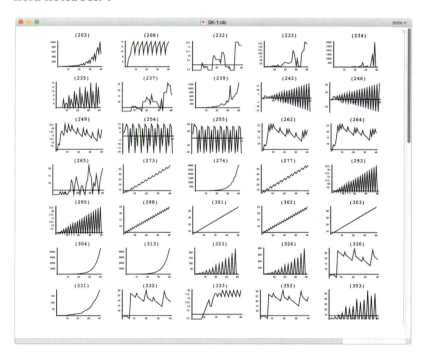

Then there are attempts to visualize combinators in the same kind of way as cellular automata:

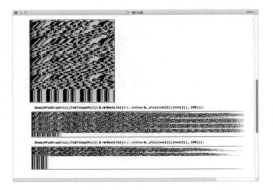

But the end result was that, yes, like Turing machines, string substitution systems and all the other systems I explored in the computational universe, combinators did exactly the same kinds of things I'd originally discovered in cellular automata. Combinators weren't just systems that could be set up to do things. Even "in the wild" they could spontaneously do very interesting and complex things.

I included a few pages on what I called "symbolic systems" (essentially lambdas) at the end of my chapter on "The World of Simple Programs" in *A New Kind of Science* (and, yes, reading particularly the notes again now, I realize there are still many more things to explore…):

Later in the book I talk specifically about Schönfinkel's combinators in connection with the threshold of computation universality. But before showing examples of what they do, I remark:

> "Originally intended as an idealized way to represent structures of functions defined in logic, combinators were actually first introduced in 1920—sixteen years before Turing machines. But although they have been investigated somewhat over the past eighty years, they have for the most part been viewed as rather obscure and irrelevant constructs."

How "irrelevant" should they be seen as being? Of course it depends on what for. As things to explore in the computational universe, cellular automata have the great advantage of allowing immediate visualization. With combinators it's a challenge to find any way to translate their behavior at all faithfully into something suitable for human perception. And since the Principle of Computational Equivalence implies that general computational features won't depend on the particulars of different systems, there's a tendency to feel that even in studying the computational universe, combinators "aren't worth the trouble".

Still, one thing that's been prominently on display with cellular automata over the past 20 or so years is the idea that any sufficiently simple system will eventually end up being a useful model for something. Mollusc pigmentation. Catalysis processes. Road traffic flow. There are simple cellular automaton models for all of these. What about combinators? Without good visualization it's harder to say "that looks like combinator behavior". And even after 100 years they're still a bit too unfamiliar. But when it comes to capturing some large-scale expression or tree behavior of some system, I won't be surprised if combinators are a good fit.

When one looks at the computational universe, one of the important ideas is "mining" it not just for programs that can serve as models for things, but also for programs that are somehow useful for some technological purpose. Yes, one can imagine specifically "compiling" some known program to combinators. But the question is whether "naturally occurring combinators" can somehow be identified as useful for some particular purpose. Could they deliver some new kind of distributed cryptographic protocol? Could they be helpful in mapping out distributed computing systems? Could they serve as a base for setting up molecular-scale computation, say with tree-like molecules? I don't know. But it will be interesting to find out. And as combinators enter their second century they provide a unique kind of "computational raw material" to mine from the computational universe.

Combinators All the Way Down?

What is the universe fundamentally made of? For a long time the assumption was that it must be described by something fundamentally mathematical. And indeed right around the time combinators were being invented the two great theories of general relativity and quantum mechanics were just developing. And in fact it seemed as if both physics and mathematics were going so well that people like David Hilbert imagined that perhaps both might be completely solved—and that there might be a mathematics-like axiomatic basis for physics that could be "mechanically explored" as he imagined mathematics could be.

But it didn't work out that way. Gödel's theorem appeared to shatter the idea of a "complete mechanical exploration" of mathematics. And while there was immense technical progress in working out the consequences of general relativity and quantum mechanics little was discovered about what might lie underneath. Computers (including things like Mathematica) were certainly useful in exploring the existing theories of physics. But physics didn't show any particular signs of being "fundamentally computational", and indeed the existing theories seemed structurally not terribly compatible with computational processes.

But as I explored the computational universe and saw just what rich and complex behavior could arise even from very simple rules, I began to wonder whether maybe, far below the level of existing physics, the universe might be fundamentally computational. I began to make specific models in which space and time were formed from an evolving network of discrete points. And I realized that some of the ideas that had arisen in the study of things like combinators and lambda calculus from the 1930s and 1940s might have direct relevance.

Like combinators (or lambda calculus) my models had the feature that they allowed many possible paths of evolution. And like combinators (or lambda calculus) at least some of my models had the remarkable feature that in some sense it didn't matter what path one took; the final result would always be the same. For combinators this "Church–Rosser" or "confluence" feature was what allowed one to have a definite fixed point that could be considered the result of a computation. In my models of the universe that doesn't just stop—things are a bit more subtle—but the generalization to what I call causal invariance is precisely what leads to relativistic invariance and the validity of general relativity.

For many years my work on fundamental physics languished—a victim of other priorities and the uphill effort of introducing new paradigms into a well-established field. But just over a year ago—with help from two very talented young physicists—I started again, with unexpectedly spectacular results.

I had never been quite satisfied with my idea of everything in the universe being represented as a particular kind of giant graph. But now I imagined that perhaps it was more like a giant symbolic expression, or, specifically, like an expression consisting of a huge collection of relations between elements—in effect, a certain kind of giant hypergraph. It was, in a way, a very combinator-like concept.

At a technical level, it's not the same as a general combinator expression: it's basically just a single layer, not a tree. And in fact that's what seems to allow the physical universe to consist of something that approximates uniform (manifold-like) space, rather than showing some kind of hierarchical tree-like structure everywhere.

But when it comes to the progression of the universe through time, it's basically just like the transformation of combinator expressions. And what's become clear is that the existence of different paths—and their ultimate equivalences—is exactly what's responsible not only for the phenomena of relativity, but also for quantum mechanics. And what's remarkable is that many of the concepts that were first discovered in the context of combinators and lambda calculus now directly inform the theory of physics. Normal forms (basically fixed points) are related to black holes where "time stops". Critical pair lemmas are related to measurement in quantum mechanics. And so on.

In practical computing, and in the creation of computational language, it was the addition of "meaningful names" to the raw structure of combinators that turned them into the powerful symbolic expressions we use. But in understanding the "data structure of the universe" we're in a sense going back to something much more like "raw combinators". Because now all those "atoms of space" that make up the universe don't have meaningful names; they're more like S's and K's in a giant combinator expression, distinct but yet all the same.

In the traditional, mathematical view of physics, there was always some sense that by "appropriately clever mathematics" it would be possible to "figure out what will happen" in any physical system. But once one imagines that physics is fundamentally computational, that's not what one can expect.

And just like combinators—with their capability for universal computation—can't in a sense be "cracked" using mathematics, so also that'll be true of the universe.

And indeed in our model that's what the progress of time is about: it's the inexorable, irreducible process of computation, associated with the repeated transformation of the symbolic expression that represents the universe.

When Hilbert first imagined that physics could be reduced to mathematics he probably thought that meant that physics could be "solved". But with Gödel's theorem—which is a reflection of universal computation—it became clear that mathematics itself couldn't just be "solved". But now in effect we have a theory that "reduces physics to mathematics", and the result of the Gödel's theorem phenomenon is something very important in our universe: it's what leads to a meaningful notion of time.

Moses Schönfinkel imagined that with combinators he was finding "building blocks for logic". And perhaps the very simplicity of what he came up with makes it almost inevitable that it wasn't just about logic: it was something much more general. Something that can represent computations. Something that has the germ of how we can represent the "machine code" of the physical universe.

It took in a sense "humanizing" combinators to make them useful for things like computational language whose very purpose is to connect with humans. But there are other places where inevitably we're dealing with something more like large-scale "combinators in the raw". Physics is one of them. But there are others. In distributed computing. And perhaps in biology, in economics and in other places.

There are specific issues of whether one's dealing with trees (like combinators), or hypergraphs (like our model of physics), or something else. But what's important is that many of the ideas—particularly around what we call multiway systems—show up with combinators. And yes, combinators often aren't the easiest places for us humans to understand the ideas in. But the remarkable fact is that they exist in combinators—and that combinators are now a century old.

I'm not sure if there'll ever be a significant area where combinators alone will be the dominant force. But combinators have—for a century—had the essence of many important ideas. Maybe as such they are at some level destined forever to be footnotes. But in a sense they are also seeds or roots—from which remarkable things have grown. And as combinators enter their second century it seems quite certain that there is still much more that will grow from them.

Его Превосходительству
Господину Ректору Императорскаго
Новороссійскаго университета

7 б

Моисея Эльева
Шейнфинкеля

Прошеніе.

Получивъ зачетное свидѣтельство
восьми семестровъ физико-матема-
тическаго факультета, математи-
ческаго отдѣленія, честь имѣю про-
сить Ваше Превосходительство
сдѣлать зависящее отъ Васъ
распоряженіе о выдачѣ мнѣ моихъ
документовъ.

Моисей Шейнфинкель.

1 апрѣля, 1910 г.

Where Did Combinators Come From? Hunting the Story of Moses Schönfinkel

December 7, 1920

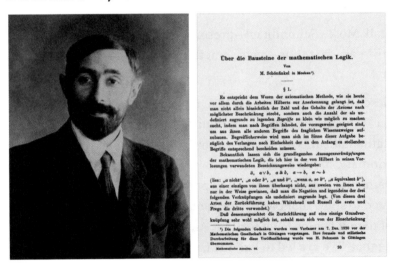

On Tuesday, December 7, 1920, the Göttingen Mathematics Society held its regular weekly meeting—at which a 32-year-old local mathematician named Moses Schönfinkel with no known previous mathematical publications gave a talk entitled "Elemente der Logik" ("Elements of Logic").

A hundred years later what was presented in that talk still seems in many ways alien and futuristic—and for most people almost irreducibly abstract. But we now realize that that talk gave the first complete formalism for what is probably the single most important idea of this past century: the idea of universal computation.

Sixteen years later would come Turing machines (and lambda calculus). But in 1920 Moses Schönfinkel presented what he called "building blocks of logic"—or what we now call "combinators"—and then proceeded to show that by appropriately combining them one could effectively define any function, or, in modern terms, that they could be used to do universal computation.

Looking back a century it's remarkable enough that Moses Schönfinkel conceptualized a formal system that could effectively capture the abstract notion of computation. And it's more remarkable still that he formulated what amounts to the idea of universal computation, and showed that his system achieved it.

But for me the most amazing thing is that not only did he invent the first complete formalism for universal computation, but his formalism is probably in some sense minimal. I've personally spent years trying to work out just how simple the structure of systems that support universal computation can be—and for example with Turing machines it took from 1936 until 2007 for us to find the minimal case.

But back in his 1920 talk Moses Schönfinkel—presenting a formalism for universal computation for the very first time—gave something that is probably already in his context minimal.

Moses Schönfinkel described the result of his 1920 talk in an 11-page paper published in 1924 entitled "Über die Bausteine der mathematischen Logik" ("On the Building Blocks of Mathematical Logic"). The paper is a model of clarity. It starts by saying that in the "axiomatic method" for mathematics it makes sense to try to keep the number of "fundamental notions" as small as possible. It reports that in 1913 Henry Sheffer managed to show that basic logic requires only one connective, that we now call NAND. But then it begins to go further. And already within a couple of paragraphs it's saying that "We are led to [an] idea, which at first glance certainly appears extremely bold". But by the end of the introduction it's reporting, with surprise, the big news: "It seems to me remarkable in the extreme that the goal we have just set can be realized... [and]; as it happens, it can be done by a reduction to three fundamental signs".

Those "three fundamental signs", of which he only really needs two, are what we now call the S and K combinators (he called them S and C). In concept they're remarkably simple, but their actual operation is in many ways brain-twistingly complex. But there they were—already a century ago—just as they are today: minimal elements for universal computation, somehow conjured up from the mind of Moses Schönfinkel.

Who Was Moses Schönfinkel?

So who was this person, who managed so long ago to see so far?

The complete known published output of Moses Schönfinkel consists of just two papers: his 1924 "On the Building Blocks of Mathematical Logic", and another, 31-page paper from 1927, coauthored with Paul Bernays, entitled "Zum *Entscheidungsproblem* der mathematischen Logik" ("On the Decision Problem of Mathematical Logic").

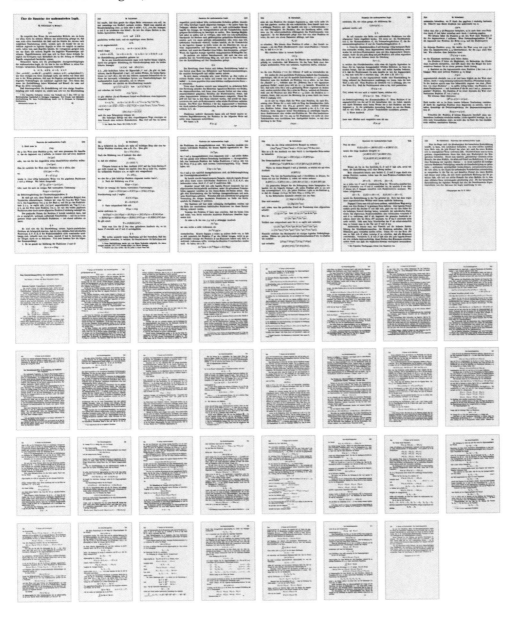

And somehow Schönfinkel has always been in the shadows—appearing at best only as a kind of footnote to a footnote. Turing machines have taken the limelight as models of computation—with combinators, hard to understand as they are, being mentioned at most only in obscure footnotes. And even within the study of combinators—often called "combinatory logic"—even as S and K have remained ubiquitous, Schönfinkel's invention of them typically garners at most a footnote.

About Schönfinkel as a person, three things are commonly said. First, that he was somehow connected with the mathematician David Hilbert in Göttingen. Second, that he spent time in a psychiatric institution. And third, that he died in poverty in Moscow, probably around 1940 or 1942.

But of course there has to be more to the story. And in recognition of the centenary of Schönfinkel's announcement of combinators, I decided to try to see what I could find out.

I don't think I've got all the answers. But it's been an interesting, if at times unsettling, trek through the Europe—and mathematics—of a century or so ago. And at the end of it I feel I've come to know and understand at least a little more about the triumph and tragedy of Moses Schönfinkel.

The Beginning of the Story

It's a strange and sad resonance with Moses Schönfinkel's life… but there's a 1953 song by Tom Lehrer about plagiarism in mathematics—where the protagonist explains his chain of intellectual theft: "I have a friend in Minsk/Who has a friend in Pinsk/Whose friend in Omsk"… "/Whose friend somehow/Is solving now/The problem in Dnepropetrovsk". Well, Dnepropetrovsk is where Moses Schönfinkel was born.

Except, confusingly, at the time it was called (after Catherine the Great or maybe her namesake saint) Ekaterinoslav (Екатеринослáв)—and it's now called Dnipro. It's one of the larger cities in Ukraine, roughly in the center of the country, about 250 miles down the river Dnieper from Kiev. And at the time when Schönfinkel was born, Ukraine was part of the Russian Empire.

So what traces are there of Moses Schönfinkel in Ekaterinoslav (AKA Dnipro) today? 132 years later it wasn't so easy to find (especially during a pandemic)… but here's a record of his birth: a certificate from the Ekaterinoslav Public Rabbi stating that entry 272 of the Birth Register for Jews from 1888 records that on September 7, 1888, a son Moses was born to the Ekaterinoslav citizen Ilya Schönfinkel and his wife Masha:

This seems straightforward enough. But immediately there's a subtlety. When exactly was Moses Schönfinkel born? What is that date? At the time the Russian Empire—which had the Russian Orthodox Church, which eschewed Pope Gregory's 1582 revision of the calendar—was still using the Julian calendar introduced by Julius Caesar. (The calendar was switched in 1918 after the Russian Revolution, although the Orthodox Church plans to go on celebrating Christmas on January 7 until 2100.) So to know a correct modern (i.e. Gregorian calendar) date of birth we have to do a conversion. And from this we'd conclude that Moses Schönfinkel was born on September 19, 1888.

But it turns out that's not the end of the story. There are several other documents associated with Schönfinkel's college years that also list his date of birth as September 7, 1888. But the state archives of the Dnepropetrovsk region contain the actual, original register from the synagogue in Ekaterinoslav. And here's entry 272—and it records the birth of Moses Schönfinkel, but on September 17, not September 7:

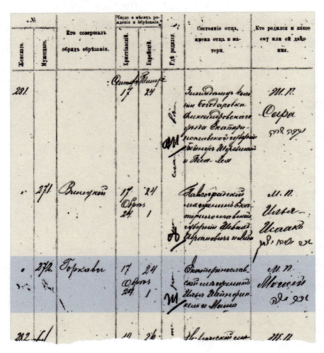

So the official certificate is wrong! Someone left a digit out. And there's a check: the Birth Register also gives the date in the Jewish calendar: 24 Tishrei—which for 1888 is the Julian date September 17. So converting to modern Gregorian form, the correct date of birth for Moses Schönfinkel is September 29, 1888.

OK, now what about his name? In Russian it's given as Моисей Шейнфинкель (or, including the patronymic, with the most common transliteration from Hebrew, Моисей Эльевич Шейнфинкель). But how should his last name be transliterated? Well, there are several possibilities. We're using Schönfinkel—but other possibilities are Sheinfinkel and Sheynfinkel—and these show up almost randomly in different documents.

What else can we learn from Moses Schönfinkel's "birth certificate"? Well, it describes his father Эльева (Ilya) as an Ekaterinoslav мещанина. But what is that word? It's often translated "bourgeoisie", but seems to have basically meant "middle-class city dweller". And in other documents from the time, Ilya Schönfinkel is described as a "merchant of the 2nd guild" (i.e. not the "top 5%" 1st guild, nor the lower 3rd guild).

Apparently, however, his fortunes improved. The 1905 "Index of Active Enterprises Incorporated in the [Russian] Empire" lists him as a "merchant of the 1st guild" and records that in 1894 he co-founded the company of "Lurie & Sheinfinkel" (with a paid-in capital of 10,000 rubles, or about $150k today) that was engaged in the grocery trade:

Lurie & Sheinfinkel seems to have had multiple wine and grocery stores. Between 1901 and 1904 its "store #2" was next to a homeopathic pharmacy in a building that probably looked at the time much like it does today:

And for store #1 there are actually contemporary photographs (note the -инкель for the end of "Schönfinkel" visible on the bottom left; this particular building was destroyed in World War II):

There seems to have been a close connection between the Schönfinkels and the Luries—who were a prominent Ekaterinoslav family involved in a variety of enterprises. Moses Schönfinkel's mother Maria (Masha) was originally a Lurie (actually, she was one of the 8 siblings of Ilya Schönfinkel's business partner Aron Lurie). Ilya Schönfinkel is listed from 1894 to 1897 as "treasurer of the Lurie Synagogue". And in 1906 Moses Schönfinkel listed his mailing address in Ekaterinoslav as Lurie House, Ostrozhnaya Square. (By 1906 that square sported an upscale park—though a century earlier it had housed a prison that was referenced in a poem by Pushkin. Now it's the site of an opera house.)

Accounts of Schönfinkel sometimes describe him as coming from a "village in Ukraine". In actuality, at the turn of the twentieth century Ekaterinoslav was a bustling metropolis, that for example had just become the third city in the whole Russian Empire to have electric trams. Schönfinkel's family also seems to have been quite well to do. Some pictures of Ekaterinoslav from the time give a sense of the environment (this building was actually the site of a Lurie candy factory):

Екатеринославъ начало XX века.
Кондитерская фабрика Лурье.
ул.Казанская (ул.К.Либкнехта, 4)

As the name "Moses" might suggest, Moses Schönfinkel was Jewish, and at the time he was born there was a large Jewish population in the southern part of Ukraine. Many Jews had come to Ekaterinoslav from Moscow, and in fact 40% of the whole population of the town was identified as Jewish.

Moses Schönfinkel went to the main high school in town (the "Ekaterinoslav classical gymnasium")—and graduated in 1906, shortly before turning 18. Here's his diploma:

 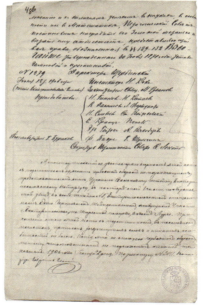

The diploma shows that he got 5/5 in all subjects—the subjects being theology, Russian, logic, Latin, Greek, mathematics, geodesy ("mathematical geography"), physics, history, geography, French, German and drawing. So, yes, he did well in high school. And in fact the diploma goes on to say: "In view of his excellent behavior and diligence and excellent success in the sciences, especially in mathematics, the Pedagogical Council decided to award him the Gold Medal..."

Going to College in Odessa

Having graduated from high school, Moses Schönfinkel wanted to go ("for purely family reasons", he said) to the University of Kiev. But being told that Ekaterinoslav was in the wrong district for that, he instead asked to enroll at Novorossiysk University in Odessa. He wrote a letter—in rather neat handwriting—to unscramble a bureaucratic issue, giving various excuses along the way:

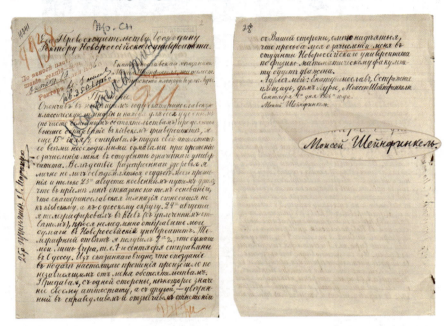

But in the fall of 1906, there he was: a student in the Faculty of Physics and Mathematics Faculty of Novorossiysk University, in the rather upscale and cosmopolitan town of Odessa, on the Black Sea.

The Imperial Novorossiya University, as it was then officially called, had been created out of an earlier institution by Tsar Alexander II in 1865. It was a distinguished university, with for example Dmitri Mendeleev (of periodic table fame) having taught there. In Soviet times it would be renamed after the discoverer of macrophages, Élie Metchnikoff (who worked there). Nowadays it is usually known as Odessa University. And conveniently, it has maintained its archives well—so that, still there, 114 years later, is Moses Schönfinkel's student file:

It's amazing how "modern" a lot of what's in it seems. First, there are documents Moses Schönfinkel sent so he could register (confirming them by telegram on September 1, 1906). There's his high-school diploma and birth certificate—and there's a document from the Ekaterinoslav City Council certifying his "citizen rank" (see above). The cover sheet also records a couple of other documents, one of which is presumably some kind of deferment of military service.

And then in the file there are two "photo cards" giving us pictures of the young Moses Schönfinkel, wearing the uniform of the Imperial Russian Army:

(These pictures actually seem to come from 1908; the style of uniform was a standard one issued after 1907; the [presumably] white collar tabs indicate the 3rd regiment of whatever division he was assigned to.)

Nowadays it would all be online, but in his physical file there is a "lecture book" listing courses (yes, every document is numbered, to correspond to a line in a central ledger):

Here are the courses Moses Schönfinkel took in his first semester in college (fall 1906):

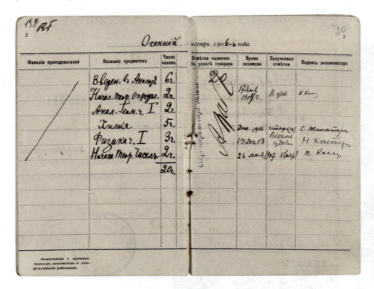

Introduction to Analysis (6 hrs), Introduction to Determinant Theory (2 hrs), Analytical Geometry 1 (2 hrs), Chemistry (5 hrs), Physics 1 (3 hrs), Elementary Number Theory (2 hrs): a total of 20 hours. Here's the bill for these courses: pretty good value at 1 ruble per course-hour, or a total of 20 rubles, which is about $300 today:

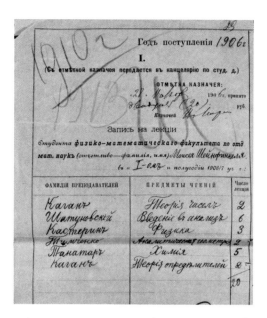

Subsequent semesters list many very familiar courses: Differential Calculus, Integrals (parts 1 and 2), and Higher Algebra, as well as "Calculus of Probabilities" (presumably probability theory) and "Determinant Theory" (essentially differently branded "linear algebra"). There are some "distribution" courses, like Astronomy (and Spherical Astronomy) and Physical Geography (or is that Geodesy?). And by 1908, there are also courses like Functions of a Complex Variable, Integro-Differential Equations (yeah, differential equations definitely pulled ahead of integral equations over the past century), Calculus of Variations and Infinite Series. And—perhaps presaging Schönfinkel's next life move— another course that makes an appearance in 1908 is German (and it's Schönfinkel's only non-science course during his whole university career).

In Schönfinkel's "lecture book" many of the courses also have names of professors listed. For example, there's "Kagan", who's listed as teaching Foundations of Geometry (as well as Higher Algebra, Determinant Theory and Integro-Differential Equations). That's Benjamin Kagan, who was then a young lecturer, but would later become a leader in differential geometry in Moscow—and also someone who studied the axiomatic foundations of geometry (as well as writing about the somewhat tragic life of Lobachevsky).

Another professor—listed as teaching Schönfinkel Introduction to Analysis and Theory of Algebraic Equation Solving—is "Shatunovsky". And (at least according to Shatunovsky's later student Sofya Yanovskaya, of whom we'll hear more later), Samuil Shatunovsky was basically Schönfinkel's undergraduate advisor.

Shatunovsky had been the 9th child of a poor Jewish family (actually) from a village in Ukraine. He was never able to enroll at a university, but for some years did manage to go to lectures by people around Pafnuty Chebyshev in Saint Petersburg. For quite a few years he then made a living as an itinerant math tutor (notably in Ekaterinoslav) but papers he wrote were eventually noticed by people at the university in Odessa, and, finally, in 1905, at the age of 46, he ended up as a lecturer at the university—where the following year he taught Schönfinkel.

Shatunovsky (who stayed in Odessa until his death in 1929) was apparently an energetic but precise lecturer. He seems to have been quite axiomatically oriented, creating axiomatic systems for geometry, algebraic fields, and notably, for order relations. (He was also quite a constructivist, opposed to the indiscriminate use of the Law of Excluded Middle.) The lectures from his Introduction to Analysis course (which Schönfinkel took in 1906) were published in 1923 (by the local publishing company Mathesis in which he and Kagan were involved).

Another of Schönfinkel's professors (from whom he took Differential Calculus and "Calculus of Probabilities") was a certain Ivan (or Jan) Śleszyński, who had worked with Karl Weierstrass on things like continued fractions, but by 1906 was in his early 50s and increasingly transitioning to working on logic. In 1911 he moved to Poland, where he sowed some of the seeds for the Polish school of mathematical logic, in 1923 writing a book called *On the Significance of Logic for Mathematics* (notably with no mention of Schönfinkel), and in 1925 one on proof theory.

It's not clear how much mathematical logic Moses Schönfinkel picked up in college, but in any case, in 1910, he was ready to graduate. Here's his final student ID (what are those pieces of string for?):

There's a certificate confirming that on April 6, 1910, Moses Schönfinkel had no books that needed returning to the library. And he sent a letter asking to graduate (with slightly-less-neat handwriting than in 1906):

The letter closes with his signature (Моисей Шейнфинкель):

Göttingen, Center of the Mathematical Universe

After Moses Schönfinkel graduated college in 1910 he probably went into four years of military service (perhaps as an engineer) in the Russian Imperial Army. World War I began on July 28, 1914—and Russia mobilized on July 30. But in one of his few pieces of good luck Moses Schönfinkel was not called up, having arrived in Göttingen, Germany, on June 1, 1914 (just four weeks before the event that would trigger World War I), to study mathematics.

Göttingen was at the time a top place for mathematics. In fact, it was sufficiently much of a "math town" that around that time postcards of local mathematicians were for sale there. And the biggest star was David Hilbert—which is who Schönfinkel went to Göttingen hoping to work with.

Hilbert had grown up in Prussia and started his career in Königsberg. His big break came in 1888 at age 26 when he got a major result in representation theory (then called "invariant theory")—using then-shocking non-constructive techniques. And it was soon after this that Felix Klein recruited Hilbert to Göttingen—where he remained for the rest of his life.

In 1900 Hilbert gave his famous address to the International Congress of Mathematicians where he first listed his (ultimately 23) problems that he thought should be important in the future of mathematics. Almost all the problems are what anyone would call "mathematical". But problem 6 has always stuck out for me: "Mathematical Treatment of the Axioms of Physics": Hilbert somehow wanted to axiomatize physics as Euclid had axiomatized geometry. And he didn't just talk about this; he spent nearly 20 years working on it. He brought in physicists to teach him, and he worked on things like gravitation theory ("Einstein–Hilbert action") and kinetic theory—and wanted for example to derive the existence of the electron from something like Maxwell's equations. (He was particularly interested in the way atomistic processes limit to continua—a problem that I now believe is deeply connected to computational irreducibility, in effect implying another appearance of undecidability, like in Hilbert's 1st, 2nd and 10th problems.)

Hilbert seemed to feel that physics was a crucial source of raw material for mathematics. But yet he developed a whole program of research based on doing mathematics in a completely formalistic way—where one just writes down axioms and somehow "mechanically" generates all true theorems from them. (He seems to have drawn some distinction between "merely mathematical" questions, and questions about physics, apparently noting—in a certain resonance with my life's work—that in the latter case "the physicist has the great calculating machine, Nature".)

In 1899 Hilbert had written down more precise and formal axioms for Euclid's geometry, and he wanted to go on and figure out how to formulate other areas of mathematics in this kind of axiomatic way. But for more than a decade he seems to have spent most of his time on physics—finally returning to questions about the foundations of mathematics around 1917, giving lectures about "logical calculus" in the winter session of 1920.

By 1920, World War I had come and gone, with comparatively little effect on mathematical life in Göttingen (the nearest battle was in Belgium 200 miles to the west). Hilbert was 58 years old, and had apparently lost quite a bit of his earlier energy (not least as a result of having contracted pernicious anemia [autoimmune vitamin B12 deficiency], whose cure was found only a few years later). But Hilbert was still a celebrity around Göttingen, and generating mathematical excitement. (Among "celebrity gossip" mentioned in a letter home by young Russian topologist Pavel Urysohn is that Hilbert was a huge fan of the gramophone, and that even at his advanced age, in the summer, he would sit in a tree to study.)

I have been able to find out almost nothing about Schönfinkel's interaction with Hilbert. However, from April to August 1920 Hilbert gave weekly lectures entitled "Problems of Mathematical Logic" which summarized the standard formalism of the field—and the official notes for those lectures were put together by Moses Schönfinkel and Paul Bernays (the "N" initial for Schönfinkel is a typo):

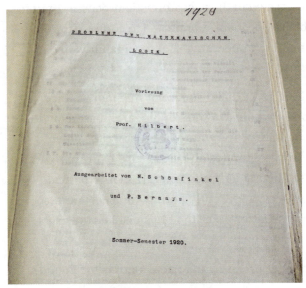

Photograph by Cem Bozşahin

A few months after these lectures came, at least from our perspective today, the highlight of Schönfinkel's time in Göttingen: the talk he gave on December 7, 1920. The venue was the weekly meeting of the Göttingen Mathematics Society, held at 6pm on Tuesdays. The society wasn't officially part of the university, but it met in the same university "Auditorium Building" that at the time housed the math institute:

The talks at the Göttingen Mathematics Society were listed in the *Annual Report of the German Mathematicians Association*:

32   Mitteilungen und Nachrichten

Mathematische Gesellschaft in Göttingen. Wintersemester 1920/21.
2. November 1920. Ferienbericht. — *9. November.* N e d e r, Trigonometrische
Reihen. — Der Vortragende sprach über ein elementares Konstruktionsver-
fahren, das bei trigonometrischen Reihen und Paaren konjugierter (d. h. Potenz-
reihen auf dem Rande des Konvergenzkreises) Beispiele mit singulären Kon-
vergenz- und Divergenzeigenschaften ergibt. Dasselbe wurde zu diesem Zweck
zuerst durch Lusin [Rendic. Pal. 32 (1911)], dann durch Steinhaus [C. R. Soc.
Sc. Varsovie 5 (1912); Bull. Ac. Cracovie 1918, S. 69] sowie (auch bei Fourier-
Reihen und Paaren konjugierter) durch den Referenten [Diss. Göttingen 1914/19;
Math. Zeitschrift 6 (1920), S. 262; Math. Annalen demnächst] angewandt. —
16. November. B e s s e l - H a g e n, Diskontinuierliche Lösungen der Variations-
rechnung. — *23. November.* R u n g e, Amerikanische Arbeiten über Sternhaufen
und die Milchstraße. — *30. November.* R ü c k l e, van der Waals jun. Über die
Erklärung der Naturgesetze auf statistisch-mechanischer Grundlage. — *7. De-
zember.* Schönfinkel, Elemente der Logik. — *14. Dezember.* B e r n a y s,
Wahrscheinlichkeit, Zeitrichtung und Kausalität. — *10. und 11. Januar 1921.*
Pezoldt, Erkenntnistheoretische Grundlagen der speziellen und allgemeinen
Relativitätstheorie. — *25. Januar.* E. N o e t h e r, Elementarteiler und allge-
meine Idealtheorie: Es wurde die Idealtheorie im Bereich aller ganzzahligen
Matrizen entwickelt und daraus eine eindeutige Zerlegung des Systems der
Elementarteiler in seine irreduziblen Bestandteile abgeleitet. [Erscheint dem-
nächst in einer Arbeit über Idealtheorie in Ringbereichen (Math. Annalen) als
Beispiel der allgemeinen Theorie.] — *1. und 8. Februar.* C o u r a n t und B e r n a y s,
Über die neuen arithmetischen Theorien von Weyl und Brouwer. — *21. und
22. Februar.* H i l b e r t, Eine neue Grundlegung des Zahlbegriffes.

There's quite a lineup. November 9, Ludwig Neder (student of Edmund Landau): "Trigonometric Series". November 16, Erich Bessel-Hagen (student of Carathéodory): "Discontinuous Solutions of Variational Problems". November 23, Carl Runge (of Runge–Kutta fame, then a Göttingen professor): "American Work on Star Clusters in the Milky Way". November 30 Gottfried Rückle (assistant of van der Waals): "Explanations of Natural Laws Using a Statistical Mechanics Basis". And then: December 7: Moses Schönfinkel, "Elements of Logic".

The next week, December 14, Paul Bernays, who worked with Hilbert and inter-acted with Schönfinkel, spoke about "Probability, the Arrow of Time and Causality" (yes, there was still a lot of interest around Hilbert in the foundations of physics). January 10+11, Joseph Petzoldt (philosopher of science): "The Epistemological Basis of Special and General Relativity". January 25, Emmy Noether (of Noether's theorem fame): "Elementary Divisors and General Ideal Theory". February 1+8, Richard Courant (of PDE etc. fame) & Paul Bernays: "About the New Arithmetic Theories of Weyl and Brouwer". February 22, David Hilbert: "On a New Basis for the Meaning of a Number" (yes, that's foundations of math).

What in detail happened at Schönfinkel's talk, or as a result of it? We don't know. But he seems to have been close enough to Hilbert that just over a year later he was in a picture taken for David Hilbert's 60th birthday on January 23, 1922:

There are all sorts of well-known mathematicians in the picture (Richard Courant, Hermann Minkowski, Edmund Landau, …) as well as some physicists (Peter Debye, Theodore von Kármán, Ludwig Prandtl, …). And there near the top left is Moses Schönfinkel, sporting a somewhat surprised expression.

For his 60th birthday Hilbert was given a photo album—with 44 pages of pictures of altogether about 200 mathematicians (and physicists). And there on page 22 is Moses Schönfinkel:

Göttingen University, Cod. Ms. D. Hilbert 754

Göttingen University, Cod. Ms. D. Hilbert 745, Bl. 22

Who are the other people on the page with him? Adolf Kratzer (1893–1983) was a student of Arnold Sommerfeld, and a "physics assistant" to Hilbert. Hermann Vermeil (1889–1959) was an assistant to Hermann Weyl, who worked on differential geometry for general relativity. Heinrich Behmann (1891–1970) was a student of Hilbert and worked on mathematical logic, and we'll encounter him again later. Finally, Carl Ludwig Siegel (1896–1981) had been a student of Landau and would become a well-known number theorist.

Problems Are Brewing

There's a lot that's still mysterious about Moses Schönfinkel's time in Göttingen. But we have one (undated) letter written by Nathan Schönfinkel, Moses's younger brother, presumably in 1921 or 1922 (yes, he romanizes his name "Scheinfinkel" rather than "Schönfinkel"):

Göttingen University, Cod. Ms. D. Hilbert 455: 9

Dear Professor!

I received a letter from Rabbi Dr. Behrens in which he wrote that my brother was in need, that he was completely malnourished. It was very difficult for me to read these lines, even more so because I cannot help my brother. I haven't received any messages or money myself for two years. Thanks to the good people where I live, I am protected from severe hardship. I am able to continue my studies. I hope to finish my PhD in 6 months. A few weeks ago I received a letter from my cousin stating that our parents and relatives are healthy. My cousin is in Kishinev (Bessarabia), now in Romania. He received the letter from our parents who live in Ekaterinoslav. Our parents want to help us but cannot do so because the postal connections are nonexistent. I hope these difficulties will not last long. My brother is helpless and impractical in this material world. He is a victim of his great love for science. Even as a 12 year old boy he loved mathematics, and all window frames and doors were painted with mathematical formulas by him. As a high school student, he devoted all his free time to mathematics. When he was studying at the university in Odessa, he was not satisfied with the knowledge there, and his striving and ideal was Göttingen and the king of mathematics, Prof. Hilbert. When he was accepted in Göttingen, he once wrote to me the following: "My dear brother, it seems to me as if I am dreaming but this is reality: I am in Göttingen, I saw Prof. Hilbert, I spoke to Prof. Hilbert." The war came and with it suffering. My brother, who is helpless, has suffered more than anyone else. But he did not write to me so as not to worry me. He has a good heart. I ask you, dear Professor, for a few months until the connections with our city are established, to help him by finding a suitable (not harmful to his health) job for him. I will be very grateful to you, dear Professor, if you will answer me.

Sincerely.

N. Scheinfinkel

We'll talk more about Nathan Schönfinkel later. But suffice it to say here that when he wrote the letter he was a physiology graduate student at the University of Bern—and he would get his PhD in 1922, and later became a professor. But the letter he wrote is probably our best single surviving source of information about the situation and personality of Moses Schönfinkel. Obviously he was a serious math enthusiast from a young age. And the letter implies that he'd wanted to work with Hilbert for some time (presumably hence the German classes in college).

It also implies that he was financially supported in Göttingen by his parents—until this was disrupted by World War I. (And we learn that his parents were OK in the Russian Revolution.) (By the way, the rabbi mentioned is probably a certain Siegfried Behrens, who left Göttingen in 1922.)

There's no record of any reply to Nathan Schönfinkel's letter from Hilbert. But at least by the time of Hilbert's 60th birthday in 1922 Moses Schönfinkel was (as we saw above) enough in the inner circle to be invited to the birthday party.

What else is there in the university archives in Göttingen about Moses Schönfinkel? There's just one document, but it's very telling:

Göttingen University, Unia GÖ, Sek. 335.55

It's dated 18 March 1924. And it's a carbon copy of a reference for Schönfinkel. It's rather cold and formal, and reads:

> "The Russian privatdozent [private lecturer] in mathematics, Mr. Scheinfinkel, is hereby certified to have worked in mathematics for ten years with Prof. Hilbert in Göttingen."

It's signed (with a stylized "S") by the "University Secretary", a certain Ludwig Gossmann, who we'll be talking about later. And it's being sent to Ms. Raissa Neuburger, at Bühlplatz 5, Bern. That address is where the Physiology Institute at the University of Bern is now, and also was in 1924. And Raissa Neuberger either was then, or soon would become, Nathan Schönfinkel's wife.

But there's one more thing, handwritten in black ink at the bottom of the document. Dated March 20, it's another note from the University Secretary. It's annotated "a.a.", i.e. *ad acta*—for the records. And in German it reads:

> *Gott sei Dank, dass Sch weg ist*

which translates in English as:

> Thank goodness Sch is gone

Hmm. So for some reason at least the university secretary was happy to see Schönfinkel go. (Or perhaps it was a German 1920s version of an HR notation: "not eligible for rehire".) But let's analyze this document in a little more detail. It says Schönfinkel worked with Hilbert for 10 years. That agrees with him having arrived in Göttingen in 1914 (which is a date we know for other reasons, as we'll see below).

But now there's a mystery. The reference describes Schönfinkel as a *"privatdozent"*. That's a definite position at a German university, with definite rules, that in 1924 one would expect to have been rigidly enforced. The basic career track was (and largely still is): first, spend 2–5 years getting a PhD. Then perhaps get recruited for a professorship, or if not, continue doing research, and write a habilitation, after which the university may issue what amounts to an official government "license to teach", making someone a privatdozent, able to give lectures. Being a privatdozent wasn't as such a paid gig. But it could be combined with a job like being an assistant to a professor—or something outside the university, like tutoring, teaching high school or working at a company.

So if Schönfinkel was a privatdozent in 1924, where is the record of his PhD, or his habilitation? To get a PhD required "formally publishing" a thesis, and printing (as in, on a printing press) at least 20 or so copies of the thesis. A habilitation was typically a substantial, published research paper. But there's absolutely no record of any of these things for Schönfinkel. And that's very surprising. Because there are detailed records for other people (like Paul Bernays) who were around at the time, and were indeed privatdozents.

And what's more the *Annual Report of the German Mathematicians Association—* which listed Schönfinkel's 1920 talk—seems to have listed mathematical goings-on in meticulous detail. Who gave what talk. Who wrote what paper. And most definitely who got a PhD, did a habilitation or became a privatdozent. (And becoming a privatdozent also required an action of the university senate, which was carefully recorded.) But going through all the annual reports of the German Mathematicians Association we find only four mentions of Schönfinkel. There's his 1920 talk, and also a 1921 talk with Paul Bernays that we'll discuss later. There's the publication of his papers in 1924 and 1927. And there's a single other entry, which says that on November 4, 1924, Richard Courant gave a report to the Göttingen Mathematical Society about a conference in Innsbruck, where Heinrich Behmann reported on "published work by M. Schönfinkel". (It describes the work as follows: "It is a continuation of Sheffer's [1913] idea of replacing the elementary operations of symbolic logic with a single one. By means of a certain function calculus, all logical statements (including the mathematical ones) are represented by three basic signs alone.")

So, it seems, the university secretary wasn't telling it straight. Schönfinkel might have worked with Hilbert for 10 years. But he wasn't a privatdozent. And actually it doesn't seem as if he had any "official status" at all.

So how do we even know that Schönfinkel was in Göttingen from 1914 to 1924? Well, he was Russian, and so in Germany he was an "alien", and as such he was required to register his address with the local police (no doubt even more so from 1914 to 1918 when Germany was, after all, at war with Russia). And the remarkable thing is that even after all these years, Schönfinkel's registration card is still right there in the municipal archives of the city of Göttingen:

Stadtarchiv Göttingen, Meldekartei

So that means we have all Schönfinkel's addresses during his time in Göttingen. Of course, there are confusions. There's yet another birthdate for Schönfinkel: September 4, 1889. Wrong year. Perhaps a wrongly done correction from the Julian calendar. Perhaps "adjusted" for some reason of military service obligations. But, in any case, the document says that Moses Schönfinkel from Ekaterinoslav arrived in Göttingen on June 1, 1914, and started living at 6 Lindenstraße (now Felix-Klein-Strasse).

He moved pretty often (11 times in 10 years), not at particularly systematic times of year. It's not clear exactly what the setup was in all these places, but at least at the end (and in another document) it lists addresses and "with Frau....", presumably indicating that he was renting a room in someone's house.

Where were all those addresses? Well, here's a map of Göttingen circa 1920, with all of them plotted (along with a red "M" for the location of the math institute):

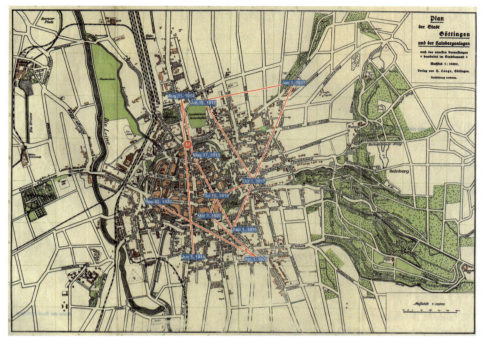

Stadtarchiv Göttingen, D 2, V a 62

The last item on the registration card says that on March 18, 1924, he departed Göttingen, and went to Moscow. And the note on the copy of the reference saying "thank goodness [he's] gone" is dated March 20, so that all ties together.

But let's come back to the reference. Who was this "University Secretary" who seems to have made up the claim that Schönfinkel was a privatdozent? It was fairly easy to find out that his name was Ludwig Gossmann. But the big surprise was to find out that the university archives in Göttingen have nearly 500 pages about him—primarily in connection with a "criminal investigation".

Here's the story. Ludwig Gossmann was born in 1878 (so he was 10 years older than Schönfinkel). He grew up in Göttingen, where his father was a janitor at the university. He finished high school but didn't go to college and started working for the local government. Then in 1906 (at age 28) he was hired by the university as its "secretary".

The position of "university secretary" was a high-level one. It reported directly to the vice-rector of the university, and was responsible for "general administrative matters" for the university, including, notably, the supervision of international students (of whom there were many, Schönfinkel being one). Ludwig Gossmann held the position of university secretary for 27 years—even while the university had a different rector (normally a distinguished academic) every year.

But Mr. Gossmann also had a sideline: he was involved in real estate. In the 1910s he started building houses (borrowing money from, among others, various university professors). And by the 1920s he had significant real estate holdings—and a business where he rented to international visitors and students at the university.

Years went by. But then, on January 24, 1933, the newspaper headline announced: "Sensational arrest: senior university official Gossmann arrested on suspicion of treason—communist revolution material [*Zersetzungsschrift*] confiscated from his apartment". It was said that perhaps it was a setup, and that he'd been targeted because he was gay (though, a year earlier, at age 54, he did marry a woman named Elfriede).

Göttingen University, Kur 3730, Sek 356 2

This was a bad time to be accused of being a communist (Hitler would become chancellor less than a week later, on January 30, 1933, in part propelled by fears of communism). Gossmann was taken to Hanover "for questioning", but was then allowed back to Göttingen "under house arrest". He'd had health problems for several years, and died of a heart attack on February 24, 1933.

But none of this really helps us understand why Gossmann would go out on a limb to falsify the reference for Schönfinkel. We can't specifically find an address match, but perhaps Schönfinkel had at least at some point been a tenant of Gossmann's. Perhaps he still owed rent. Perhaps he was just difficult in dealing with the university administration. It's not clear. It's also not clear why the reference Gossmann wrote was sent to Schönfinkel's brother in Bern, even though Schönfinkel himself was going to Moscow. Or why it wasn't just handed to Schönfinkel before he left Göttingen.

The 1924 Paper

Whatever was going on with Schönfinkel in Göttingen in 1924, we know one thing for sure: it was then that he published his remarkable paper about what are now called combinators. Let's talk in a bit more detail about the paper—though the technicalities I'm discussing in part two.

First, there's some timing. At the end of the paper, it says it was received by the journal on March 15, 1924, i.e. just three days before the date of Ludwig Gossmann's reference for Schönfinkel. And then at the top of the paper, there's something else: under Schönfinkel's name it says "in Moskau", i.e. at least as far as the journal was concerned, Schönfinkel was in Moscow, Russia, at the time the article was published:

Über die Bausteine der mathematischen Logik.

Von

M. Schönfinkel in Moskau[1]).

There's also a footnote on the first page of the paper:

[1]) Die folgenden Gedanken wurden vom Verfasser am 7. Dez. 1920 vor der Mathematischen Gesellschaft in Göttingen vorgetragen. Ihre formale und stilistische Durcharbeitung für diese Veröffentlichung wurde von H. Behmann in Göttingen übernommen.

"The following thoughts were presented by the author to the Mathematical Society in Göttingen on December 7, 1920. Their formal and stylistic processing for this publication was done by H. Behmann in Göttingen."

The paper itself is written in a nice, clear and mathematically mature way. Its big result (as I've discussed elsewhere) is the introduction of what would later be called combinators: two abstract constructs from which arbitrary functions and computations can be built up. Schönfinkel names one of them S, after the German word "*Schmelzen*" for "fusion". The other has become known as K, although Schönfinkel calls it C, even though the German word for "constancy" (which is what would naturally describe it) is "*Konstantheit*", which starts with a K.

The paper ends with three paragraphs, footnoted with "The considerations that follow are the editor's" (i.e. Behmann's). They're not as clear as the rest of the paper, and contain a confused mistake.

The main part of the paper is "just math" (or computation, or whatever). But here's the page where S and K (called C here) are first used:

---

312   M. Schönfinkel.

Da y willkürlich ist, können wir dafür ein beliebiges Ding oder eine beliebige Funktion einsetzen, also z. B. Cx. Dies gibt:

$$Ix = (Cx)(Cx).$$

Nach der Erklärung von S bedeutet dies aber:

$$SCCx,$$

so daß wir erhalten:

$$I = SCC. \text{ }^3)$$

Übrigens kommt es in dem Ausdruck SCC auf das letzte Zeichen C gar nicht einmal an. Setzen wir nämlich oben für y nicht Cx, sondern die willkürliche Funktion φx, so ergibt sich entsprechend:

$$I = SC\varphi,$$

wo also für φ jede beliebige Funktion eingesetzt werden kann $^4)$.

    2. Nach der Erklärung von Z ist

$$Zfgx = f(gx).$$

Weiter ist vermöge der bereits verwendeten Umformungen

$$f(gx) = (Cfx)(gx) = S(Cf)gx = (CSf)(Cf)gx.$$

Verschmelzung nach f ergibt:

$$S(CS)Cfgx,$$

also

$$Z = S(CS)C.$$

    3. Ganz entsprechend läßt sich

$$Tfyx = fxy$$

weiter umformen in:

$$fx(Cyx) = (fx)(Cyx) = Sf(Cy)x = (Sf)(Cy)x = Z(Sf)Cyx$$
$$= ZZSfCyx = (ZZSf)Cyx = (ZZSf)(CCf)yx = S(ZZS)(CC)fyx.$$

Es gilt somit:

$$T = S(ZZS)(CC).$$

    Setzt man hier für Z den oben gefundenen Ausdruck ein, so ist damit T ebenfalls auf C und S zurückgeführt.

§ 5.

    Wir wollen nunmehr unsere Ergebnisse auf den besonderen Fall des Logikkalküls anwenden, in welchem die Grundelemente die Individuen und

---

$^3)$ Diese Zurückführung wurde mir von Herrn Boskowitz mitgeteilt, die etwas weniger einfache $(SC)(CC)$ bereits früher von Herrn Bernays.
$^4)$ Freilich nur eine solche, die für jedes x einen Sinn hat.

And now there's something more people-oriented: a footnote to the combinator equation $I = SCC$ saying "This reduction was communicated to me by Mr. Boskowitz; some time before that, Mr. Bernays had called the somewhat less simple one $(SC)(CC)$ to my attention." In other words, even if nothing else, Schönfinkel had talked to Boskowitz and Bernays about what he was doing.

OK, so we've got three people—in addition to David Hilbert—somehow connected to Moses Schönfinkel.

Let's start with Heinrich Behmann—the person footnoted as "processing" Schönfinkel's paper for publication:

He was born in Bremen, Germany, in 1891, making him a couple of years younger than Schönfinkel. He arrived in Göttingen as a student in 1911, and by 1914 was giving a talk about Whitehead and Russell's *Principia Mathematica* (which had been published in 1910). When World War I started he volunteered for military service, and in 1915 he was wounded in action in Poland (receiving an Iron Cross)—but in 1916 he was back in Göttingen studying under Hilbert, and in 1918 he wrote his PhD thesis on "The Antinomy of the Transfinite Number and Its Resolution by the Theory of Russell and Whitehead" (i.e. using the idea of types to deal with paradoxes associated with infinity).

Behmann continued in the standard academic track (i.e. what Schönfinkel apparently didn't do)—and in 1921 he got his habilitation with the thesis "Contributions to the Algebra of Logic, in Particular to the *Entscheidungsproblem* [Decision Problem]". There'd been other decision problems discussed before, but Behmann said what he meant was a "procedure [giving] complete instructions for determining whether a [logical or mathematical] assertion is true or false by a deterministic calculation after finitely many steps". And, yes, Alan Turing's 1936 paper "On Computable Numbers, with an Application to the *Entscheidungsproblem*" was what finally established that the halting problem, and therefore the *Entscheidungsproblem*, was undecidable. Curiously, in principle, there should have been enough in Schönfinkel's paper that this could have been figured out back in 1921 if Behmann or others had been thinking about it in the right way (which might have been difficult before Gödel's work).

So what happened to Behmann? He continued to work on mathematical logic and the philosophy of mathematics. After his habilitation in 1921 he became a privat-dozent at Göttingen (with a job as an assistant in the applied math institute), and then in 1925 got a professorship in Halle in applied math—though having been an active member of the Nazi Party since 1937, lost this professorship in 1945 and became a librarian. He died in 1970.

(By the way, even though in 1920 "PM" [*Principia Mathematica*] was hot—and Behmann was promoting it—Schönfinkel had what in my opinion was the good taste to not explicitly mention it in his paper, referring only to Hilbert's much-less-muddy ideas about the formalization of mathematics.)

OK, so what about Boskowitz, credited in the footnote with having discovered the classic combinator result $I = SKK$? That was Alfred Boskowitz, in 1920 a 23-year-old Jewish student at Göttingen, who came from Budapest, Hungary, and worked with Paul Bernays on set theory. Boskowitz is notable for having contributed far more corrections (nearly 200) to *Principia Mathematica* than anyone else, and being acknowledged (along with Behmann) in a footnote in the (1925–27) second edition. (This edition also gives a reference to Schönwinkel's [*sic*] paper at the end of a list of 14 "other contributions to mathematical logic" since the first edition.) In the mid-1920s Boskowitz returned to Budapest. In 1936 he wrote to Behmann that anti-Jewish sentiment there made him concerned for his safety. There's one more known communication from him in 1942, then no further trace.

The third person mentioned in Schönfinkel's paper is Paul Bernays, who ended up living a long and productive life, mostly in Switzerland. But we'll come to him later.

So where was Schönfinkel's paper published? It was in a journal called *Mathematische Annalen* (*Annals of Mathematics*)—probably the top math journal of the time. Here's its rather swank masthead, with quite a collection of famous names (including physicists like Einstein, Born and Sommerfeld):

The "instructions to contributors" on the inside cover of each issue had a statement from the "Editorial Office" about not changing things at the proof stage because "according to a calculation they [cost] 6% of the price of a volume". The instructions then go on to tell people to submit papers to the editors—at their various home addresses (it seems David Hilbert lived just down the street from Felix Klein…):

Die   MATHEMATISCHEN ANNALEN

erscheinen in Heften, von denen je vier einen Band von etwa 20 Bogen bilden. Sie sind durch jede Buchhandlung sowie durch die Verlagsbuchhandlung zu beziehen. Die Mitglieder der Deutschen Mathematiker-Vereinigung haben Anspruch auf einen Vorzugspreis.

Die Verfasser erhalten von Abhandlungen bis zu 24 Seiten Umfang 100 Sonderabdrucke, von größeren Arbeiten 50 Sonderabdrucke kostenfrei, weitere gegen Berechnung.

Geschäftsführender Redakteur ist

O. Blumenthal, Aachen, Rütscherstraße 38.

Alle Korrektursendungen sind an ihn zu richten.

Für die „Mathematischen Annalen" bestimmte Manuskripte können bei jedem der unten verzeichneten Redaktionsmitglieder eingereicht werden:

L. Bieberbach, Berlin-Schmargendorf, Marienbaderstraße 9,
O. Blumenthal, Aachen, Rütscherstraße 38,
H. Bohr, Kopenhagen, St. Hans Torv 32,
M. Born, Göttingen, Planckstraße 21,
L. E. J. Brouwer, Laren (Nordholland),
C. Carathéodory, München, Universität,
R. Courant, Göttingen, Nikolausbergerweg 5,
W. v. Dyck, München, Hildegardstraße 5,
A. Einstein, Berlin-Wilmersdorf, Haberlandstraße 5,
D. Hilbert, Göttingen, Wilhelm-Weber-Straße 29,
O. Hölder, Leipzig, Schenkendorfstraße 8,
Th. v. Kármán, Aachen, Nizzaallee 41,
F. Klein, Göttingen, Wilhelm-Weber-Straße 3,
C. Neumann, Leipzig, Querstraße 10—12,
A. Sommerfeld, München, Leopoldstraße 87.

Here's the complete table of contents for the volume in which Schönfinkel's paper appears:

There are a variety of famous names here. But particularly notable for our purposes are Aleksandr Khintchine (of Khinchin constant fame) and the topologists Pavel Alexandroff and Pavel Urysohn, who were all from Moscow State University, and who are all indicated, like Schönfinkel, as being "in Moscow".

There's a little bit of timing information here. Schönfinkel's paper was indicated as having been received by the journal on March 15, 1924. The "thank goodness [he's] gone [from Göttingen]" comment is dated March 20. Meanwhile, the actual issue of the journal with Schönfinkel's article (number 3 of 4) was published September 15, with table of contents:

But note the ominous † next to Urysohn's name. Turns out his fatal swimming accident was August 17, so—notwithstanding their admonitions—the journal must have added the † quite quickly at the proof stage.

The "1927" Paper

Beyond his 1924 paper on combinators, there's only one other known piece of published output from Moses Schönfinkel: a paper coauthored with Paul Bernays "On the Decision Problem of Mathematical Logic":

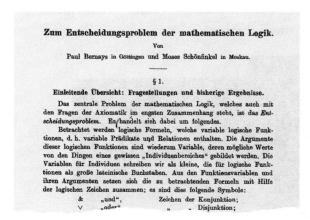

It's actually much more widely cited than Schönfinkel's 1924 combinator paper, but it's vastly less visionary and ultimately much less significant; it's really about a technical point in mathematical logic.

About halfway through the paper it has a note:

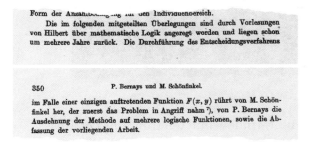

"The following thoughts were inspired by Hilbert's lectures on mathematical logic and date back several years. The decision procedure for a single function $F(x, y)$ was derived by M. Schönfinkel, who first tackled the problem; P. Bernays extended the method to several logical functions, and also wrote the current paper."

The paper was submitted on March 24, 1927. But in the records of the German Mathematicians Association we find a listing of another talk at the Göttingen Mathematical Society: December 6, 1921, P. Bernays and M. Schönfinkel, "Das *Entscheidungsproblem* im Logikkalkul". So the paper had a long gestation period, and (as the note in the paper suggests) it basically seems to have fallen to Bernays to get it written, quite likely with little or no communication with Schönfinkel.

So what else do we know about it? Well, remarkably enough, the Bernays archive contains two notebooks (the paper kind!) by Moses Schönfinkel that are basically an early draft of the paper (with the title already being the same as it finally was, but with Schönfinkel alone listed as the author):

ETH Zurich, Bernays Archive, Hs. 974: 282

These notebooks are basically our best window into the front lines of Moses Schönfinkel's work. They aren't dated as such, but at the end of the second notebook there's a byline of sorts, that lists his street address in Göttingen—and we know he lived at that address from September 1922 until March 1924:

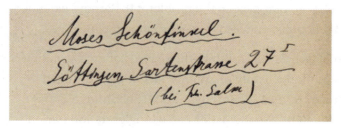

OK, so what's in the notebooks? The first page might indicate that the notebooks were originally intended for a different purpose. It's just a timetable of lectures:

"Hilbert lectures: Monday: Mathematical foundations of quantum theory; Thursday: Hilbert–Bernays: Foundations of arithmetic; Saturday: Hilbert: Knowledge and mathematical thinking". (There's also a slightly unreadable note that seems to say "Hoppe. 6–8... electricity", perhaps referring to Edmund Hoppe, who taught physics in Göttingen, and wrote a history of electricity.)

But then we're into 15 pages (plus 6 in the other notebook) of content, written in essentially perfect German, but with lots of parentheticals of different possible word choices:

The final paper as coauthored with Bernays begins:

> "The central problem of mathematical logic, which is also closely connected to its axiomatic foundations, is the decision problem [*Entscheidungsproblem*]. And it deals with the following. We have logical formulas which contain logic functions, predicates, ..."

Schönfinkel's version begins considerably more philosophically (here with a little editing for clarity):

> "Generality has always been the main goal—the ideal of the mathematician. Generality in the solution, in the method, in the concept and formulation of the theorem, in the problem and question. This tendency is even more pronounced and clearer with modern mathematicians than with earlier ones, and reaches its high point in the work of Hilbert and Ms. Noether. Such an ideal finds its most extreme expression when one faces the problem of 'solving all problems'—at least all mathematical problems, because everything else after is easy, as soon as this 'Gordian Knot' is cut (because the world is written in 'mathematical letters' according to Hilbert).

> In just the previous century mathematicians would have been extremely skeptical and even averse to such fantasies... But today's mathematician has already been trained and tested in the formal achievements of modern mathematics and Hilbert's axiomatics, and nowadays one has the courage and the boldness to dare to touch this question as well. We owe to mathematical logic the fact that we are able to have such a question at all.

> From Leibniz's bold conjectures, the great logician-mathematicians went step by step in pursuit of this goal, in the systematic structure of mathematical logic: Boole (discoverer of the logical calculus), (Bolzano?), Ernst Schröder, Frege, Peano, Ms. Ladd-Franklin, the two Peirces, Sheffer, Whitehead, Couturat, Huntington, Padoa, Shatunovsky, Sleshinsky, Kagan, Poretsky, Löwenheim, Skolem, ... and their numerous students, collaborators and contemporaries... until in 1910–1914 'the system' by Bertrand Russell and Whitehead appeared—the famous 'Principia Mathematica'—a mighty titanic work, a large system. Finally came our knowledge of logic from Hilbert's lectures on (the algebra of) logic (-calculus) and, following on from this, the groundbreaking work of Hilbert's students: Bernays and Behmann.

The investigations of all these scholars and researchers have led (in no uncertain terms) to the fact that it has become clear that actual mathematics represents a branch of logic. … This emerges most clearly from the treatment and conception of mathematical logic that Hilbert has given. And now, thanks to Hilbert's approach, we can (satisfactorily) formulate the great decision problem of mathematical logic."

We learn quite a bit about Schönfinkel from this. Perhaps the most obvious thing is that he was a serious fan of Hilbert and his approach to mathematics (with a definite shout-out to "Ms. Noether"). It's also interesting that he refers to Bernays and Behmann as "students" of Hilbert. That's pretty much correct for Behmann. But Bernays (as we'll see soon) was more an assistant or colleague of Hilbert's than a student.

It gives interesting context to see Schönfinkel rattle off a sequence of contributors to what he saw as the modern view of mathematical logic. He begins—quite rightly I think—mentioning "Leibniz's bold conjectures". He's not sure whether Bernard Bolzano fits (and neither am I). Then he lists Schröder, Frege and Peano—all pretty standard choices, involved in building up the formal structure of mathematical logic.

Next he mentions Christine Ladd-Franklin. At least these days, she's not particularly well known, but she had been a mathematical logic student of Charles Peirce, and in 1881 she'd written a paper about the "Algebra of Logic" which included a truth table, a solid 40 years before Post or Wittgenstein. (In 1891 she had also worked in Göttingen on color vision with the experimental psychologist Georg Müller—who was still there in 1921.) It's notable that Schönfinkel mentions

Ladd-Franklin ahead of the father-and-son Peirces. Next we see Sheffer, who Schönfinkel quotes in connection with NAND in his combinator paper. (No doubt unbeknownst to Schönfinkel, Henry Sheffer—who spent most of his life in the US—was also born in Ukraine ["near Odessa", his documents said], and was also Jewish, and was just 6 years older than Schönfinkel.) I'm guessing Schönfinkel mentions Whitehead next in connection with universal algebra, rather than his later collaboration with Russell.

Next comes Louis Couturat, who frankly wouldn't have made my list for mathematical logic, but was another "algebra of logic" person, as well as a Leibniz fan, and developer of the Ido language offshoot from Esperanto. Huntington was involved in the axiomatization of Boolean algebra; Padoa was connected to Peano's program. Shatunovsky, Sleshinsky and Kagan were all professors of Schönfinkel's in Odessa (as mentioned above), concerned in various ways with foundations of mathematics. Platon Poretsky I must say I had never heard of before; he seems to have done fairly technical work on propositional logic. And finally Schönfinkel lists Löwenheim and Skolem, both of whom are well known in mathematical logic today.

I consider it rather wonderful that Schönfinkel refers to Whitehead and Russell's *Principia Mathematica* as a "titanic work" (*Titanenwerk*). The showy and "overconfident" *Titanic* had come to grief on its iceberg in 1912, somehow reminiscent of *Principia Mathematica*, eventually coming to grief on Gödel's theorem.

At first it might just seem charming—particularly in view of his brother's comment that "[Moses] is helpless and impractical in this material world"—to see Schönfinkel talk about how after one's solved all mathematical problems, then solving *all problems* will be easy, explaining that, after all, Hilbert has said that "the world is written in 'mathematical letters'". He says that in the previous century mathematicians wouldn't have seriously considered "solving everything", but now, because of progress in mathematical logic, "one has the courage and the boldness to dare to touch this question".

It's very easy to see this as naive and unworldly—the writing of someone who knew only about mathematics. But though he didn't have the right way to express it, Schönfinkel was actually onto something, and something very big. He talks at the beginning of his piece about generality, and about how recent advances in mathematical logic embolden one to pursue it. And in a sense he was very right about this. Because mathematical logic—through work like his—is what led us to the modern conception of computation, which really is successful in "talking about everything". Of course, after Schönfinkel's time we learned about Gödel's theorem and computational irreducibility, which tell us that even though we may be able to talk about everything, we can never expect to "solve every problem" about everything.

But back to Schönfinkel's life and times. The remainder of Schönfinkel's notebooks give the technical details of his solution to a particular case of the decision problem. Bernays obviously worked through these, adding more examples as well as some generalization. And Bernays cut out Schönfinkel's philosophical introduction, no doubt on the (probably correct) assumption that it would seem too airy-fairy for the paper's intended technical audience.

So who was Paul Bernays? Here's a picture of him from 1928:

Bernays was almost exactly the same age as Schönfinkel (he was born on October 17, 1888—in London, where there was no calendar issue to worry about). He came from an international business family, was a Swiss citizen and grew up in Paris and Berlin. He studied math, physics and philosophy with a distinguished roster of professors in Berlin and Göttingen, getting his PhD in 1912 with a thesis on analytic number theory.

After his PhD he went to the University of Zurich, where he wrote a habilitation (on complex analysis), and became a privatdozent (yes, with the usual documentation, that can still be found), and an assistant to Ernst Zermelo (of ZFC set theory fame). But in 1917 Hilbert visited Zurich and soon recruited Bernays to return to Göttingen. In Göttingen, for apparently bureaucratic reasons, Bernays wrote a second habilitation, this time on the axiomatic structure of *Principia Mathematica* (again, all the documentation can still be found). Bernays was also hired to work as a "foundations of math assistant" to Hilbert. And it was presumably in that capacity that he—along with Moses Schönfinkel—wrote the notes for Hilbert's 1920 course on mathematical logic.

Unlike Schönfinkel, Bernays followed a fairly standard—and successful—academic track. He became a professor in Göttingen in 1922, staying there until he was dismissed (because of partially Jewish ancestry) in 1933—after which he moved back to Zurich, where he stayed and worked very productively, mostly in mathematical logic (von Neumann–Bernays–Gödel set theory, etc.), until he died in 1977.

Back when he was in Göttingen one of the things Bernays did with Hilbert was to produce the two-volume classic *Grundlagen der Mathematik* (*Foundations of Mathematics*). So did the *Grundlagen* mention Schönfinkel? It has one mention of the Bernays–Schönfinkel paper, but no direct mention of combinators. However, there is one curious footnote:

$$\bar{A} \to A,$$

¹ Vgl. S. 78.
² Ein System von Ausgangsformeln, das zur Ableitung aller identisch wahren Implikationsformeln ausreicht, hat zuerst M. Schönfinkel aufgestellt. A. Tarski hat erkannt, daß als Ausgangsformeln für diesen Bereich schon die drei Formeln I 1), I 3) und

$$((A \to B) \to C) \to ((A \to C) \to C)$$

genügen. Hier kann noch die dritte Formel durch die obengenannte einfachere ersetzt werden. Von M. Wajsberg und J. Lukasiewicz wurden verschiedene Formeln gefunden, die jede für sich schon als einzige Ausgangsformel zur Ableitung aller identisch wahren Implikationsformeln ausreichen. Vgl. hierzu den schon genannten Bericht von Lukasiewicz und Tarski: „Untersuchungen über den Aussagenkalkül" (C. R. Soc. Sci. Varsovie Bd. 23. Warschau 1930).

This starts "A system of axioms that is sufficient to derive all true implicational formulas was first set up by M. Schönfinkel...", then goes on to discuss work by Alfred Tarski. So do we have evidence of something else Schönfinkel worked on? Probably.

In ordinary logic, one starts from an axiom system that gives relations, say about AND, OR and NOT. But, as Sheffer established in 1910, it's also possible to give an axiom system purely in terms of NAND (and, yes, I'm proud to say that I found the very simplest such axiom system in 2000). Well, it's also possible to use other bases for logic. And this footnote is about using IMPLIES as the basis. Actually, it's implicational calculus, which isn't as strong as ordinary logic, in the sense that it only lets you prove some of the theorems. But there's a question again: what are the possible axioms for implicational calculus?

Well, it seems that Schönfinkel found a possible set of such axioms, though we're not told what they were; only that Tarski later found a simpler set. (And, yes, I looked for the simpler axiom systems for implicational calculus in 2000, but didn't find any.) So again we see Schönfinkel in effect trying to explore the lowest-level foundations of mathematical logic, though we don't know any details.

So what other interactions did Bernays have with Schönfinkel? There seems to be no other information in Bernays's archives. But I have been able to get a tiny bit more information. In a strange chain of connections, someone who's worked on Mathematica and Wolfram Language since 1987 is Roman Maeder. And Roman's thesis advisor (at ETH Zurich) was Erwin Engeler—who was a student of Paul Bernays. Engeler (who is now in his 90s) worked for many years on combinators, so of course I had to ask him what Bernays might have told him about Schönfinkel. He told me he recalled only two conversations. He told me he had the impression that Bernays found Schönfinkel a difficult person. He also said he believed that the last time Bernays saw Schönfinkel it was in Berlin, and that Schönfinkel was somehow in difficult circumstances. Any such meeting in Berlin would have had to be before 1933. But try as we might to track it down, we haven't succeeded.

To Moscow and Beyond...

In the space of three days in March 1924 Moses Schönfinkel—by then 35 years old—got his paper on combinators submitted to *Mathematische Annalen*, got a reference for himself sent out, and left for Moscow. But why did he go to Moscow? We simply don't know.

A few things are clear, though. First, it wasn't difficult to get to Moscow from Göttingen at that time; there was pretty much a direct train there. Second, Schönfinkel presumably had a valid Russian passport (and, one assumes, didn't have any difficulties from not having served in the Russian military during World War I).

One also knows that there was a fair amount of intellectual exchange and travel between Göttingen and Moscow. The very same volume of *Mathematische Annalen* in which Schönfinkel's paper was published has three (out of 19) authors in addition to Schönfinkel listed as being in Moscow: Pavel Alexandroff, Pavel Urysohn and Aleksandr Khintchine. Interestingly, all of these people were at Moscow State University.

And we know there was more exchange with that university. Nikolai Luzin, for example, got his PhD in Göttingen in 1915, and went on to be a leader in mathematics at Moscow State University (until he was effectively dismissed by Stalin in 1936). And we know that for example in 1930, Andrei Kolmogorov, having just graduated from Moscow State University, came to visit Hilbert.

Did Schönfinkel go to Moscow State University? We don't know (though we haven't yet been able to access any archives that may be there).

Did Schönfinkel go to Moscow because he was interested in communism? Again, we don't know. It's not uncommon to find mathematicians ideologically sympathetic to at least the theory of communism. But communism doesn't seem to have particularly been a thing in the mathematics or general university community in Göttingen. And indeed when Ludwig Gossmann was arrested in 1933, investigations of who he might have recruited into communism didn't find anything of substance.

Still, as I'll discuss later, there is a tenuous reason to think that Schönfinkel might have had some connection to Leon Trotsky's circle, so perhaps that had something to do with him going to Moscow—though it would have been a bad time to be involved with Trotsky, since by 1925 he was already out of favor with Stalin.

A final theory is that Schönfinkel might have had relatives in Moscow; at least it looks as if some of his Lurie cousins ended up there.

But realistically we don't know. And beyond the bylines on the journals, we don't really have any documentary evidence that Schönfinkel was in Moscow. However, there is one more data point, from November 1927 (8 months after the submission of Schönfinkel's paper with Bernays). Pavel Alexandroff was visiting Princeton University, and when Haskell Curry (who we'll meet later) asked him about Schönfinkel he was apparently told that "Schönfinkel has... gone insane and is now in a sanatorium & will probably not be able to work any more."

Ugh! What happened? Once again, we don't know. Schönfinkel doesn't seem to have ever been "in a sanatorium" while he was in Göttingen; after all, we have all his addresses, and none of them were sanatoria. Maybe there's a hint of something in Schönfinkel's brother's letter to Hilbert. But are we really sure that Schönfinkel actually suffered from mental illness? There's a bunch of hearsay that says he did. But then it's a common claim that logicians who do highly abstract work are prone to mental illness (and, well, yes, there are a disappointingly large number of historical examples).

Mental illness wasn't handled very well in the 1920s. Hilbert's only child, his son Franz (who was about five years younger than Schönfinkel), suffered from mental illness, and after a delusional episode that ended up with him in a clinic, David Hilbert simply said "From now on I have to consider myself as someone who does not have a son". In Moscow in the 1920s—despite some political rhetoric—conditions in psychiatric institutions were probably quite poor, and there was for example quite a bit of use of primitive shock therapy (though not yet electroshock). It's notable, by the way, that Curry reports that Alexandroff described Schönfinkel as being "in a sanatorium". But while at that time the word "sanatorium" was being used in the US as a better term for "insane asylum", in Russia it still had more the meaning of a place for a rest cure. So this still doesn't tell us if Schönfinkel was in fact "institutionalized"—or just "resting". (By the way, if there was mental illness involved, another connection for Schönfinkel that doesn't seem to have been made is that Paul Bernays's first cousin once removed was Martha Bernays, wife of Sigmund Freud.)

Whether or not he was mentally ill, what would it have been like for Schönfinkel in what was then the Soviet Union in the 1920s? One thing is that in the Soviet system, everyone was supposed to have a job. So Schönfinkel was presumably employed doing something—though we have no idea what. Schönfinkel had presumably been at least somewhat involved with the synagogue in Göttingen (which is how the rabbi there knew to tell his brother he was in bad shape). There was a large and growing Jewish population in Moscow in the 1920s, complete with things like Yiddish newspapers. But by the mid 1930s it was no longer so comfortable to be Jewish in Moscow, and Jewish cultural organizations were being shut down.

By the way, in the unlikely event that Schönfinkel was involved with Trotsky, there could have been trouble even by 1925, and certainly by 1929. And it's notable that it was a common tactic for Stalin (and others) to claim that their various opponents were "insane".

So what else do we know about Schönfinkel in Moscow? It's said that he died there in 1940 or 1942, aged 52–54. Conditions in Moscow wouldn't have been good then; the so-called Battle of Moscow occurred in the winter of 1941. And there are various stories told about Schönfinkel's situation at that time.

The closest to a primary source seems to be a summary of mathematical logic in the Soviet Union, written by Sofya Yanovskaya in 1948. Yanovskaya was born in 1896 (so 8 years after Schönfinkel), and grew up in Odessa. She attended the same university there as Schönfinkel, studying mathematics, though arrived five years after Schönfinkel graduated. She had many of the same professors as Schönfinkel, and, probably like Schönfinkel, was particularly influenced by Shatunovsky. When the Russian Revolution happened, Yanovskaya went "all in", becoming a serious party operative, but eventually began to teach, first at the Institute of Red Professors, and then from 1925 at Moscow State University—where she became a major figure in mathematical logic, and was eventually awarded the Order of Lenin.

One might perhaps have thought that mathematical logic would be pretty much immune to political issues. But the founders of communism had talked about mathematics, and there was a complex debate about the relationship between Marxist–Leninist ideology and formal ideas in mathematics, notably the Law of Excluded Middle. Sofya Yanovskaya was deeply involved, initially in trying to "bring mathematics to heel", but later in defending it as a discipline, as well as in editing Karl Marx's mathematical writings.

It's not clear to what extent her historical writings were censored or influenced by party considerations, but they certainly contain lots of good information, and in 1948 she wrote a paragraph about Schönfinkel:

> **7.** Существенную роль в дальнейшем развитии математической логики сыграла работа М. И. Шейнфинкеля [1]. Этот блестящий ученик С. О. Шатуновского, к сожалению, рано выбыл из строя. (Заболев душевно, М. И. Шейнфинкель умер в Москве в 1942 г.). Работа, о которой идёт речь, была выполнена им в 1920 г., но опубликована только в 1924 г. в литературном оформлении Бемана. Непосредственной целью

"The work of M. I. Sheinfinkel played a substantial role in the further development of mathematical logic. This brilliant student of S. O. Shatunovsky, unfortunately, left us early. (After getting mentally ill [заболев душевно], M. I. Sheinfinkel passed away in Moscow in 1942.) He did the work mentioned here in 1920, but only published it in 1924, edited by Behmann."

Unless she was hiding things, this quote doesn't make it sound as if Yanovskaya knew much about Schönfinkel. (By the way, her own son was apparently severely mentally ill.) A student of Jean van Heijenoort (who we'll encounter later) named Irving Anellis did apparently in the 1990s ask a student of Yanovskaya's whether Yanovskaya had known Schönfinkel. Apparently he responded that unfortunately nobody had thought to ask her that question before she died in 1966.

What else do we know? Nothing substantial. The most extensively embellished story I've seen about Schönfinkel appears in an anonymous comment on the talk page for the Wikipedia entry about Schönfinkel:

> "William Hatcher, while spending time in St Petersburg during the 1990s, was told by Soviet mathematicians that Schönfinkel died in wretched poverty, having no job and but one room in a collective apartment. After his death, the rough ordinary people who shared his apartment burned his manuscripts for fuel (WWII was raging). The few Soviet mathematicians around 1940 who had any discussions with Schönfinkel later said that those mss reinvented a great deal of 20th century mathematical logic. Schönfinkel had no way of accessing the work of Turing, Church, and Tarski, but had derived their results for himself. Stalin did not order Schönfinkel shot or deported to Siberia, but blame for Schönfinkel's death and inability to publish in his final years can be placed on Stalin's doorstep. 202.36.179.65 06:50, 25 February 2006 (UTC)"

William Hatcher was a mathematician and philosopher who wrote extensively about the Bahá'í Faith and did indeed spend time at the Steklov Institute of Mathematics in Saint Petersburg in the 1990s—and mentioned Schönfinkel's technical work in his writings. People I've asked at the Steklov Institute do remember Hatcher, but don't know anything about what it's claimed he was told about Schönfinkel. (Hatcher died in 2005, and I haven't been successful at getting any material from his archives.)

So are there any other leads? I did notice that the IP address that originated the Wikipedia comment is registered to the University of Canterbury in New Zealand. So I asked people there and in the New Zealand foundations of math scene. But despite a few "maybe so-and-so wrote that" ideas, nobody shed any light.

OK, so what about at least a death certificate for Schönfinkel? Well, there's some evidence that the registry office in Moscow has one. But they tell us that in Russia only direct relatives can access death certificates....

Other Schönfinkels...

So far as we know, Moses Schönfinkel never married, and didn't have children. But he did have a brother, Nathan, who we encountered earlier in connection with the letter he wrote about Moses to David Hilbert. And in fact we know quite a bit about Nathan Scheinfinkel (as he normally styled himself). Here's a biographical summary from 1932:

Scheinfinkel, Nathan, Dr. med., P.-D., Neufeldstr. 5a, B e r n (geb. 13. IX. 93 i n Ekaterinoslaw, Ukraine). Nat.: Ausländer. — Stud. Gymn. Ekaterinoslaw, Univ. Bern. 20 med. Dr.-Examen; seit 22 Assistent a. Berner Physiolog. Institut (Ausbildung unt. Prof. Asher nach d. biochem. u. biophysikal. Richtung im Geiste d. modernen Forschung.) Seit 29 P.-D. f. Physiologie a. d. Univ. Bern. — V.: Sämtl. Arbeiten beschäftigen sich mit den Problemen: Ermüdung, Sauerstoffmangel, Herzarbeit u. Wirkungsweise d. Herznerven (ersch. i. d. Zeitschr. f. Biologie).

Deutsches Biographisches Archiv, II 1137, 103

The basic story is that he was about five years younger than Moses, and went to study medicine at the University of Bern in Switzerland in April 1914 (i.e. just before World War I began). He got his MD in 1920, then got his PhD on "Gas Exchange and Metamorphosis of Amphibian Larvae after Feeding on the Thyroid Gland or Substances Containing Iodine" in 1922. He did subsequent research on the electrochemistry of the nervous system, and in 1929 became a privatdozent—with official "license to teach" documentation:

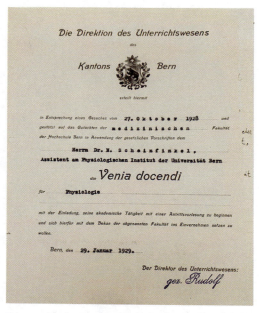

Staatsarchiv des Kantons Bern, BB IIIb 557 Scheinfinkel N.

(In a piece of bizarre small-worldness, my grandfather, Max Wolfram, also got a PhD in the physiology [veterinary medicine] department at the University of Bern [studying the function of the thymus gland], though that was in 1909, and presumably he had left before Nathan Scheinfinkel arrived.)

But in any case, Nathan Scheinfinkel stayed at Bern, eventually becoming a professor, and publishing extensively, including in English. He became a Swiss citizen in 1932, with the official notice stating:

> "Scheinfinkel, Nathan. Son of Ilia Gerschow and Mascha [born] Lurie, born in Yekaterinoslav, Russia, September 13, 1893 (old style). Doctor of medicine, residing in Bern, Neufeldstrasse 5a, husband of Raissa [born] Neuburger."

In 1947, however, he moved to become a founding professor in a new medical school in Ankara, Turkey. (Note that Turkey, like Switzerland, had been neutral in World War II.) In 1958 he moved again, this time to found the Institute of Physiology at Ege University in Izmir, Turkey, and then at age 67, in 1961, he retired and returned to Switzerland.

Did Nathan Scheinfinkel have children (whose descendents, at least, might know something about "Uncle Moses")? It doesn't seem so. We tracked down Nuran Harirî, now an emeritus professor, but in the 1950s a young physiology resident at Ege University responsible for translating Nathan Scheinfinkel's lectures into Turkish. She said that Nathan Scheinfinkel was at that point living in campus housing with his wife, but she never heard mention of any children, or indeed of any other family members.

What about any other siblings? Amazingly, looking through handwritten birth records from Ekaterinoslav, we found one! Debora Schönfinkel, born December 22, 1889 (i.e. January 3, 1890, in the modern calendar):

So Moses Schönfinkel had a younger sister, as well as a younger brother. And we even know that his sister graduated from high school in June 1907. But we don't know anything else about her, or about other siblings. We know that Schönfinkel's mother died in 1936, at the age of 74.

Might there have been other Schönfinkel relatives in Ekaterinoslav? Perhaps, but it's unlikely they survived World War II—because in one of those shocking and tragic pieces of history, over a four-day period in February 1942 almost the whole Jewish population of 30,000 was killed.

Could there be other Schönfinkels elsewhere? The name is not common, but it does show up (with various spellings and transliterations), both before and after Moses Schönfinkel. There's a Scheinfinkel Russian revolutionary buried in the Kremlin Wall; there was a Lovers of Zion delegate Scheinfinkel from Ekaterinoslav. There was a Benjamin Scheinfinkel in New York City in the 1940s; a Shlomo Scheinfinkel in Haifa in the 1930s. There was even a certain curiously named Bas Saul Haskell Scheinfinkel born in 1875. But despite quite a bit of effort, I've been unable to locate any living relative of Moses Schönfinkel. At least so far.

Haskell Curry

What happened with combinators after Schönfinkel published his 1924 paper? Initially, so far as one can tell, nothing. That is, until Haskell Curry found Schönfinkel's paper in the library at Princeton University in November 1927—and launched into a lifetime of work on combinators.

Who was Haskell Curry? And why did he know to care about Schönfinkel's paper?

Haskell Brooks Curry was born on September 12, 1900, in a small town near Boston, MA. His parents were both elocution educators, who by the time Haskell Curry was born were running the School of Expression (which had evolved from his mother's Boston-based School of Elocution and Expression). (Many years later, the School of Expression would evolve into Curry College in Waltham, Massachusetts—which happens to be where for several years we held our Wolfram Summer School, often noting the "coincidence" of names when combinators came up.)

Haskell Curry went to college at Harvard, graduating in mathematics in 1920. After a couple of years doing electrical engineering, he went back to Harvard, initially working with Percy Bridgman, who was primarily an experimental physicist, but was writing a philosophy of science book entitled *The Logic of Modern Physics*. And perhaps through this Curry got introduced to Whitehead and Russell's *Principia Mathematica*.

But in any case, there's a note in his archive about *Principia Mathematica* dated May 20, 1922:

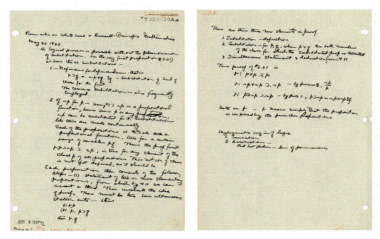

Haskell P. Curry papers, PSUA 222, Special Collections Library, Pennsylvania State University

Curry seems—perhaps like an electrical engineer or a "pre-programmer"—to have been very interested in the actual process of mathematical logic, starting his notes with: "No logical process is possible without the phenomenon of substitution." He continued, trying to break down the process of substitution.

But then his notes end, more philosophically, and perhaps with "expression" influence: "Phylogenetic origin of logic: 1. Sensation; 2. Association: Red hot poker–law of permanence".

At Harvard Curry started working with George Birkhoff towards a PhD on differential equations. But by 1927–8 he had decided to switch to logic, and was spending a year as an instructor at Princeton. And it was there—in November 1927—that he found Schönfinkel's paper. Preserved in his archives are the notes he made:

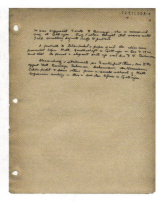

Haskell P. Curry papers, PSUA 222, Special Collections Library, Pennsylvania State University

At the top there's a date stamp of November 28, 1927. Then Curry writes: "This paper anticipates much of what I have done"—then launches into a formal summary of Schönfinkel's paper (charmingly using *f@x* to indicate function application—just as we do in Wolfram Language, except his is left associative…).

He ends his "report" with "In criticism I might say that no formal development have been undertaken in the above. Equality is taken intuitively and such things as universality, and proofs of identity are shown on the principle that if for every *z*, *x@z* : *y@z* then *x=y*…."

But then there's another piece:

"On discovery of this paper I saw Prof. Veblen. Schönfinkel's paper said 'in Moskau'. Accordingly we sought out Paul Alexandroff. The latter says Schönfinkel has since gone insane and is now in a sanatorium & will probably not be able to work any more. The paper was written with help of Paul Bernays and Behman [*sic*]; who would presumably be the only people in the world who would write on that subject."

What was the backstory to this? Oswald Veblen was a math professor at Princeton who had worked on the axiomatization of geometry and was by then working on topology. Pavel Alexandroff (who we encountered earlier) was visiting from Moscow State University for the year, working on topology with Hopf, Lefschetz, Veblen and Alexander. I'm not quite sure why Curry thought Bernays and Behmann "would be the only people in the world who would write on that subject"; I don't see how he could have known.

Curry continues: "It was suggested I write to Bernays, who is *außerord*. prof. [long-term lecturer] at Göttingen." But then he adds—in depressingly familiar academic form: "Prof. Veblen thought it unwise until I had something definite ready to publish."

"A footnote to Schönfinkel's paper said the ideas were presented before Math Gesellschaft in Göttingen on Dec. 7, 1920 and that its formal and elegant [*sic*] write up was due to H. Behman". "Elegant" is a peculiar translation of "*stilistische*" that probably gives Behmann too much credit; a more obvious translation might be "stylistic".

Curry continues: "Alexandroff's statements, as I interpret them, are to the effect that Bernays, Behman, Ackermann, von Neumann, Schönfinkel & some others form a small school of math logicians working on this & similar topics in Göttingen."

And so it was that Curry resolved to study in Göttingen, and do his PhD in logic there. But before he left for Göttingen, Curry wrote a paper (published in 1929):

An Analysis of Logical Substitution.
By H. B. Curry.

*Contents.**

I. Preliminary Discussion of the Nature of Mathematical Logic.
II. Logical Substitution; its Relation to a Combinatory Problem.
III. Solution of the Combinatory Problem.

I.

Mathematical Logic has been defined as an application of the formal methods of mathematics to the domain of Logic.† Logic, on the other hand, is the analysis and criticism of thought.‡ In accordance with those definitions, the essential purpose of mathematical logic is the construction of an abstract (or strictly formalized) theory, such that when its fundamental notions are properly interpreted, there ensues an analysis of those universal principles in accordance with which valid thinking goes on. The term **analysis** here means that a certain rather complicated body of knowledge is exhibited as derivable from a much simpler body assumed at the beginning. Evidently the simpler this initial knowledge, and the more explicitly and carefully it is set forth, the more profound and satisfactory is the analysis concerned.

Already there's something interesting in the table of contents: the use of the word "combinatory", which, yes, in Curry's care is going to turn into "combinator".

The paper starts off reading a bit like a student essay, and one's not encouraged by a footnote a few pages in:

"In writing the foregoing account I have naturally made use of any ideas I may have gleaned from reading the literature. The writings of Hilbert are fundamental in this connection. I hope that I have added clearness to certain points where the existing treatments are obscure." ["Clearness" not "clarity"?]

Then, towards the end of the "Preliminary Discussion" is this:

> Now although the rule of inference, stated above, is simple enough, yet in all current mathematical logics there exist rules which are highly complex. The presence of these complex rules raises the question whether it is possible to formulate a theory which is—1) adequate for the whole of logic, 2) based on a finite number of primitive ideas, postulates, and rules, the last of the same order of complexity as the rule of inference. I believe that it is; indeed steps in that direction have already been taken.* As a preliminary to treating this general problem, I shall discuss in the rest of this paper a special one connected with it; viz., the analysis of the process of substitution. The latter process is one of those complicated rules which occur in practically every logical theory to-day.

And the footnote says: "See the paper of Schönfinkel cited below". It's (so far as I know) the first-ever citation to Schönfinkel's paper!

On the next page Curry starts to give details. Curry starts talking about substitution, then says (in an echo of modern symbolic language design) this relates to the idea of "transformation of functions":

> A notion closely related to substitution is that of transformation of functions. Suppose we regard a function as having inherent in its definition a certain order of its variables. Then permuting these variables in any way, or making two or more of them alike, will produce new functions related to the old; let us call them transforms of the original function, and the operations by which they are produced transformations. If we number the variables consecutively 1, 2, 3, $\cdots$, then the transforms for a function of two variables will be—
>
> $$\phi(1, 2), \phi(2, 1), \phi(1, 1).$$
>
> For three variables there will be 13 transforms, for four variables 75, for five variables 541, etc. It is clear that the process of substituting a series of constants in an arbitrary manner (such that the total number of entities counting repetitions is n) into the original function is equivalent to the substitution of the same entities in a prescribed manner (viz., the first entity into the place of the first variable, the second into that of the second, etc.) into one of the transforms. The study of substitution is thus to a certain degree equivalent to the study of these transformations.
>
> An important step toward the analysis of this situation was made by M. Schönfinkel.* Starting, apparently, from the fact that every logical formula is a combination of constants—the variables being only apparent—he shows that neither the notions of propositional function (of various orders)
>
> _____
> * "Ueber die Bausteine der mathematischen Logik," _Mathematische Annalen_, Vol. 92 (1924), pp. 305-316.

At first he's off talking about all the various combinatorial arrangements of variables, etc. But then he introduces Schönfinkel—and starts trying to explain in a formal way what Schönfinkel did. And even though he says he's talking about what one assumes is structural substitution, he seems very concerned about what equality means, and how Schönfinkel didn't quite define that. (And, of course, in the end, with universal computation, undecidability, etc. we know that the definition of equality wasn't really accessible in the 1920s.)

By the next page, here we are, S and K (Curry renamed Schönfinkel's C):

CURRY: *An Analysis of Logical Substitution.* 371

III. POSTULATES.

None.

IV. RULES.

0. If x and y are entities, then (xy) shall be an entity.

1. $(=)$ shall have the properties of identity. These properties may be specified by a few simple rules; but in this treatment we shall not go into that detail. We shall treat $(=)$ as if it were precisely the intuitive relation of equality.

2. If x and y are any entities, then

$$Kxy = x$$

3. If x, y, z are entities, then

$$Sxyz = xz(yz)$$

4. If X and Y are combinations of S and K, and if there exists an integer n such that by application of the preceding rules we can formally reduce the expressions $Xx_1 x_2 \cdots x_n$ and $Yx_1 x_2 \cdots x_n$ to combinations of $x_1 x_2 \cdots x_n$ which have the same structure, then $X = Y$.

If the above primitive frame were a part of a general theory of logic, the term entity would include not only the various combinations of S and K, but all the notions of logic as well. In the sequel we shall accordingly speak of the application of combinations S and K to various logical notions, and of the resulting notions to each other, just as if these notions had been adjoined to the above frame.

The *raison d'être* of the theory based on this frame is the following fact: Let $x_1, x_2, \cdots, x_n$ be any n entities, and X any combination of them con-

At first he's imagining that the combinators have to be applied to something (i.e. $f[x]$ not just f). But by the next page he comes around to what Schönfinkel was doing in looking at "pure combinators":

class is one which has particular reference to logical substitution.

To begin with, we make the following definitions (the first three were made by Schönfinkel) : *

$$I = SKK$$
$$B = S(KS)K$$
$$C_1 = S(BBS)(KK)$$
$$C_2 = BC_1 ; C_3 = BC_2 ; \cdots \text{ etc.}$$
$$W = SS(SK)$$

then the reader may verify that whenever $x_0, x_1, x_2 \cdots$ are entities

$$Ix_0 = x_0$$
$$Bx_0 x_1 x_2 = x_0(x_1 x_2)$$
$$C_1 x_0 x_1 x_2 = x_0 x_2 x_1$$
$$C_2 x_0 x_1 x_2 x_3 = x_0 x_1 x_3 x_2$$
$$\cdots \cdots \cdots \cdots$$
$$Wx_0 x_1 = x_0 x_1 x_1$$

* We use B and C respectively in place of Schönfinkel's Z and T. Nothing corresponding to W or $C_2, C_3, \cdots$ is defined by him.

The rest of the paper is basically concerned with setting up combinators that can successively represent permutations—and it certainly would have been much easier if Curry had had a computer (and one could imagine minimal "combinator sorters" like minimal sorting networks):

After writing this paper, Curry went to Göttingen—where he worked with Bernays. I must say that I'm curious what Bernays said to Curry about Schönfinkel (was it more than to Erwin Engeler?), and whether other people around Göttingen even remembered Schönfinkel, who by then had been gone for more than four years. In 1928, travel in Europe was open enough that Curry should have had no trouble going, for example, to Moscow, but there's no evidence he made any effort to reach out to Schönfinkel. But in any case, in Göttingen he worked on combinators, and over the course of a year produced his first official paper on "combinatory logic":

Strangely, the paper was published in an American journal—as the only paper not in English in that volume. The paper is more straightforward, and in many ways more "Schönfinkel like". But it was just the first of many papers that Curry wrote about combinators over the course of nearly 50 years.

Curry was particularly concerned with the "mathematicization" of combinators, finding and fixing problems with axioms invented for them, connecting to other formalisms (notably Church's lambda calculus), and generally trying to prove theorems about what combinators do. But more than that, Curry spread the word about combinators far and wide. And before long most people viewed him as "Mr. Combinator", with Schönfinkel at most a footnote.

In 1958, when Haskell Curry and Robert Feys wrote their book on *Combinatory Logic*, there's a historical footnote—that gives the impression that Curry "almost" had Schönfinkel's ideas before he saw Schönfinkel's paper in 1927:

1. Historical statement

The material in this chapter came mostly from [GKL], [PKR], and Rosser [MLV]. These have been supplemented from notes not previously published, and from specific suggestions made by various persons, notably Church and Bernays.

The combinators B, C, I, K, and S were introduced by Schönfinkel [BML] (see § 0D). This paper is still valuable for the discussion of the intuitive meaning of these combinators and the motivation for introducing them. (There is an error in the supplementary statement by Behmann at the end of the paper, in that it is not always possible to remove parentheses by B alone; Behmann has written that this error was pointed out by Boscovitch at an early date.)

The earliest work of Curry (till the fall of 1927), which was done without knowledge of the work of Schönfinkel, used B, C, W, and I as primitive combinators. (These symbols were suggested by English, rather than German, names; 'B' by 'substitution', since 'S' might be used for several other purposes such as 'sum', 'successor', etc.; and 'W' by a natural association of this letter with repetition.) When Schönfinkel was discovered in a literature search, K was added to the theory at once;

but S was regarded as a mere technicality until the development axiomatic theories in the 1940's (see § 6S1).

The combinators B^m, C_n, K_n, W_n were introduced in [GKL]; The combinators Z_n were suggested, at least in principle, by Ch The connection between combinators and arithmetic was elabora and Rosser (cf. § 0C); Rosser's [MLV] was especially significant i connection. The connections of B with products and powers was

I have to say that I don't think that's a correct impression. What Schönfinkel did was much more singular than that. It's plausible to think that others (and particularly Curry) could have had the idea that there could be a way to go "below the operations of mathematical logic" and find more fundamental building blocks based on understanding things like the process of substitution. But the actuality of how Schönfinkel did it is something quite different—and something quite unique.

And when one sees Schönfinkel's *S* combinator: what mind could have come up with such a thing? Even Curry says he didn't really understand the significance of the *S* combinator until the 1940s.

I suppose if one's just thinking of combinatory logic as a formal system with a certain general structure then it might not seem to matter that things as simple as *S* and *K* can be the ultimate building blocks. But the whole point of what Schönfinkel was trying to do (as the title of his paper says) was to find the "building blocks of logic". And the fact that he was able to do it—especially in terms of things as simple as *S* and *K*—was a great and unique achievement. And not something that (despite all the good he did for combinators) Curry did.

Schönfinkel Rediscovered

In the decade or so after Schönfinkel's paper appeared, Curry occasionally referenced it, as did Church and a few other closely connected people. But soon Schönfinkel's paper—and Schönfinkel himself—disappeared completely from view, and standard databases list no citations.

But in 1967 Schönfinkel's paper was seen again—now even translated into English. The venue was a book called *From Frege to Gödel: A Source Book in Mathematical Logic, 1879–1931*. And there, sandwiched between von Neumann on transfinite numbers and Hilbert on "the infinite", is Schönfinkel's paper, in English, with a couple of pages of introduction by Willard Van Orman Quine. (And indeed it was from this book that I myself first became aware of Schönfinkel and his work.)

But how did Schönfinkel's paper get into the book? And do we learn anything about Schönfinkel from its appearance there? Maybe. The person who put the book together was a certain Jean van Heijenoort, who himself had a colorful history. Born in 1912, he grew up mostly in France, and went to college to study mathematics—but soon became obsessed with communism, and in 1932 left to spend what ended up being nearly ten years working as a kind of combination PR person and bodyguard for Leon Trotsky, initially in Turkey but eventually in Mexico. Having married an American, van Heijenoort moved to New York City, eventually enrolling in a math PhD program, and becoming a professor doing mathematical logic (though with some colorful papers along the way, with titles like "The Algebra of Revolution").

Why is this relevant? Well, the question is: how did van Heijenoort know about Schönfinkel? Perhaps it was just through careful scholarship. But just maybe it was through Trotsky. There's no real evidence, although it is known that during his time in Mexico, Trotsky did request a copy of *Principia Mathematica* (or was it his "PR person"?). But at least if there was a Trotsky connection it could help explain Schönfinkel's strange move to Moscow. But in the end we just don't know.

What Should We Make of Schönfinkel?

When one reads about the history of science, there's a great tendency to get the impression that big ideas come suddenly to people. But my historical research—and my personal experience—suggest that that's essentially never what happens. Instead, there's usually a period of many years in which some methodology or conceptual framework gradually develops, and only then can the great idea emerge.

So with Schönfinkel it's extremely frustrating that we just can't see that long period of development. The records we have just tell us that Schönfinkel announced combinators on December 7, 1920. But how long had he been working towards them? We just don't know.

On the face of it, his paper seems simple—the kind of thing that could have been dashed off in a few weeks. But I think it's much more likely that it was the result of a decade of development—of which, through foibles of history, we now have no trace.

Yes, what Schönfinkel finally came up with is simple to explain. But to get to it, he had to cut through a whole thicket of technicality—and see the essence of what lay beneath. My life as a computational language designer has often involved doing very much this same kind of thing. And at the end of it, what you come up with may seem in retrospect "obvious". But to get there often requires a lot of hard intellectual work.

And in a sense what Schönfinkel did was the most impressive possible version of this. There were no computers. There was no ambient knowledge of computation as a concept. Yet Schönfinkel managed to come up with a system that captures the core of those ideas. And while he didn't quite have the language to describe it, I think he did have a sense of what he was doing—and the significance it could have.

What was the personal environment in which Schönfinkel did all this? We just don't know. We know he was in Göttingen. We don't think he was involved in any particularly official way with the university. Most likely he was just someone who was "around". Clearly he had some interaction with people like Hilbert and Bernays. But we don't know how much. And we don't really know if they ever thought they understood what Schönfinkel was doing.

Even when Curry picked up the idea of combinators—and did so much with it—I don't think he really saw the essence of what Schönfinkel was trying to do. Combinators and Schönfinkel are a strange episode in intellectual history. A seed sown far ahead of its time by a person who left surprisingly few traces, and about whom we know personally so little.

But much as combinators represent a way of getting at the essence of computation, perhaps in combinators we have the essence of Moses Schönfinkel: years of a life compressed to two "signs" (as he would call them) S and K. And maybe if the operation we now call currying needs a symbol we should be using the "sha" character Ш from the beginning of Schönfinkel's name to remind us of a person about whom we know so little, but who planted a seed that gave us so much.

Thanks

Many people and organizations have helped in doing research and providing material for this piece. Thanks particularly to Hatem Elshatlawy (fieldwork in Göttingen, etc.), Erwin Engeler (first-person history), Unal Goktas (Turkish material), Vitaliy Kaurov (locating Ukraine + Russia material), Anna & Oleg Marichev (interpreting old Russian handwriting), Nik Murzin (fieldwork in Moscow), Eila Stiegler (German translations), Michael Trott (interpreting German). Thanks also for input from Henk Barendregt, Semih Baskan, Metin Baştuğ, Cem Bozşahin, Jason Cawley, Jack Copeland, Nuran Harirî, Ersin Koylu, Alexander Kuzichev, Yuri Matiyasevich, Roman Maeder, Volker Peckhaus, Jonathan Seldin, Vladimir Shalack, Matthew Szudzik, Christian Thiel, Richard Zach. Particular thanks to the following archives and staff: Berlin State Library [Gabriele Kaiser], Bern University Archive [Niklaus Bütikofer], ETHZ (Bernays) Archive [Flavia Lanini, Johannes Wahl], Göttingen City Archive [Lena Uffelmann], Göttingen University [Katarzyna Chmielewska, Bärbel Mund, Petra Vintrová, Dietlind Willer].

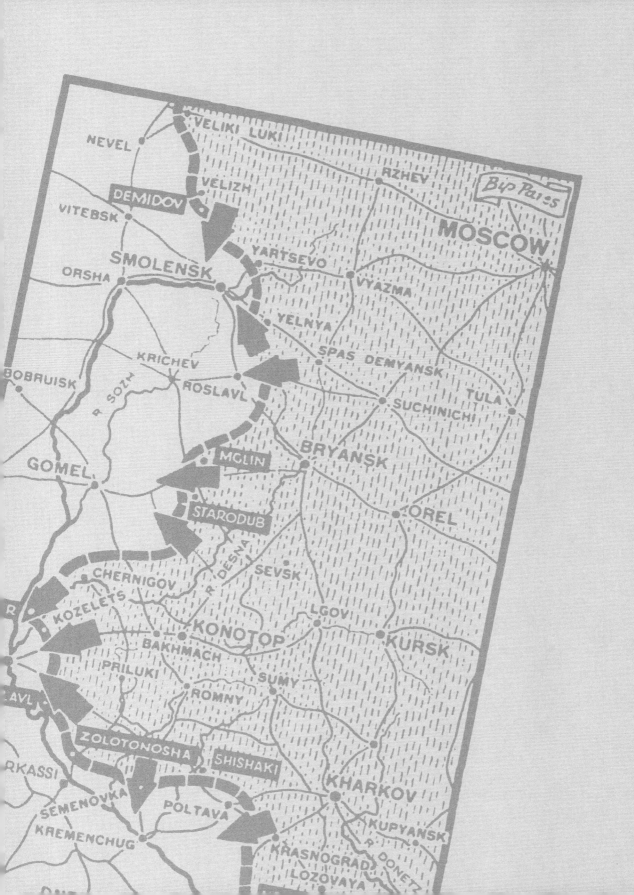

A Little Closer to Finding What Became of Moses Schönfinkel, Inventor of Combinators

For most big ideas in recorded intellectual history one can answer the question: "What became of the person who originated it?" But late last year I tried to answer that for Moses Schönfinkel, who sowed a seed for what's probably the single biggest idea of the past century: abstract computation and its universality.

I managed to find out quite a lot about Moses Schönfinkel. But I couldn't figure out what became of him. Still, I kept on digging. And it turns out I was able to find out more. So here's an update....

To recap a bit: Moses Schönfinkel was born in 1888 in Ekaterinoslav (now Dnipro) in what's now Ukraine. He went to college in Odessa, and then in 1914 went to Göttingen to work with David Hilbert. He didn't publish anything, but on December 7, 1920—at the age of 32—he gave a lecture entitled "Elemente der Logik" ("Elements of Logic") that introduced what are now called combinators, the first complete formalism for what we'd now call abstract computation. Then on March 18, 1924, with a paper based on his lecture just submitted for publication, he left for Moscow. And basically vanished.

It's said that he had mental health issues, and that he died in poverty in Moscow in 1940 or 1942. But we have no concrete evidence for either of these claims.

When I was researching this last year, I found out that Moses Schönfinkel had a younger brother Nathan Scheinfinkel (yes, he used a different transliteration of the Russian Шейнфинкель) who became a physiology professor at Bern in Switzerland, and later in Turkey. Late in the process, I also found out that Moses Schönfinkel had a younger sister Debora, who we could tell graduated from high school in 1907.

Moses Schönfinkel came from a Jewish merchant family, and his mother came from a quite prominent family. I suspected that there might be other siblings (Moses's mother came from a family of 8). And the first "new find" was that, yes, there were indeed two additional younger brothers. Here are the recordings of their births now to be found in the State Archives of the Dnipropetrovsk (i.e. Ekaterinoslav) Region:

So the complete complement of Шейнфинкель/Schönfinkel/Scheinfinkel children was (including birth dates both in their original Julian calendar form, and in their modern Gregorian form, and graduation dates in modern form):

	born (original calendar)	born (modern calendar)	graduated high school
Moses (Моисей)	September 17, 1888	September 29, 1888	June 26, 1906
Debora (Дебора)	December 22, 1889	January 3, 1890	June 25, 1907
Nathan (Натан)	September 13, 1893	September 25, 1893	June 16, 1913
Israel (Израиль)	December 5, 1894	December 17, 1894	June 16, 1913
Gregory (Григорий)	April 30, 1899	May 12, 1899	? 1917

And having failed to find out more about Moses Schönfinkel directly, plan B was to investigate his siblings.

I had already found out a fair amount about Nathan. He was married, and lived at least well into the 1960s, eventually returning to Switzerland. And most likely he had no children.

Debora we could find no trace of after her high-school graduation (we looked for marriage records, but they're not readily available for what we assume is the relevant time period).

By the way, rather surprisingly, we found nice (alphabetically ordered), printed class lists from the high-school graduations (apparently these were distributed to higher-education institutions across the Russian Empire so anyone could verify "graduation status", and were deposited in the archives of the education district, where they've now remained for more than a century):

(We can't find any particular trace of the 36 other students in the same group as Moses.)

OK, so what about the "newly found siblings", Israel and Gregory? Well, here we had a bit more luck.

For Israel we found these somewhat strange traces:

They are World War I hospital admission records from January and December 1916. Apparently Israel was a private in the 2nd Finnish Regiment (which—despite its name—by then didn't have any Finns in it, and in 1916 was part of the Russian 7th Army pushing west in southern Ukraine in the effort to retake Galicia). And the documents we have show that twice he ended up in a hospital in Pavlohrad (only about 40 miles from Ekaterinoslav, though in the opposite direction from where the 7th Army was) with some kind of (presumably not life-threatening) hernia-like problem.

But unfortunately, that's it. No more trace of Israel.

OK, what about the "baby brother", Gregory, 11 years younger than Moses? Well, he shows up in World War II records. We found four documents:

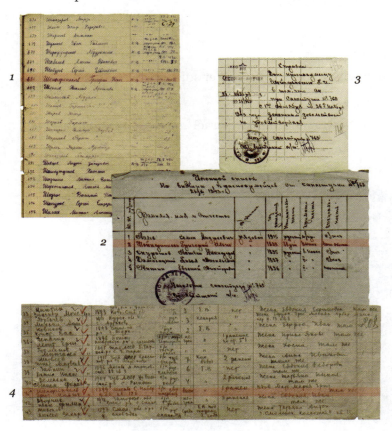

Document #4 contains something interesting: an address for Gregory in 1944—in Moscow. Remember that Moses went to Moscow in 1924. And one of my speculations was that this was the result of some family connection there. Well, at least 20 years later (and probably also much earlier, as we'll see), his brother Gregory was in Moscow. So perhaps that's why Moses went there in 1924.

OK, but what story do these World War II documents tell about Gregory? Document #1 tells us that on July 27, 1943, Gregory arrived at the military unit designated 15 зсп 44 зсбр (15 ZSP 44 ZSBR) at transit point (i.e. basically "military address") 215 азсп 61A (215 AZSP 61A). It also tells us that he had the rank of private in the Red Army.

Sometime soon thereafter he was transferred to unit 206 ZSP. But unfortunately he didn't last long in the field. Around October 1, 1943, he was wounded (later, we learn he has "one wound"), and—as document #2 tells us—he was one of 5 people picked up by hospital train #762 (at transit point 206 зсп ЗапФ). On November 26, 1943, document #3 records that he was discharged from the hospital train (specifically, the document explains that he's not getting paid for the time he was on the hospital train). And, finally, document #4 records that on February 18, 1944—presumably after a period of assessment of his condition—he's discharged from the military altogether, returning to an address in Moscow.

OK, so first some military points. When Gregory arrived in the army in July 1943 he was assigned (as a reserve or "replacement") to the 44th Rifle Brigade 44 зсбр) in the 15th Rifle Division (15 зсп) in the 61st Army (61A)—presumably as part of reinforcements brought in after some heavy Soviet losses. Later he was transferred to the 206th Rifle Division in the 47th Army, which is where he was when he was wounded around October 1, 1943.

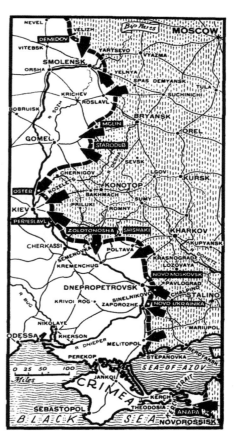

What was the general military situation then? In the summer of 1943 the major story was that the Soviets were trying to push the Germans back west, with the front pretty much along the Dnieper River in Ukraine—which, curiously enough, flows right through the middle of Ekaterinoslav. In late September 1943, here's how newspapers were presenting things ▶

But military history being what it is, there's much more detailed information available. Here's a modern map showing troop movements involving the 47th Army in late September 1943:

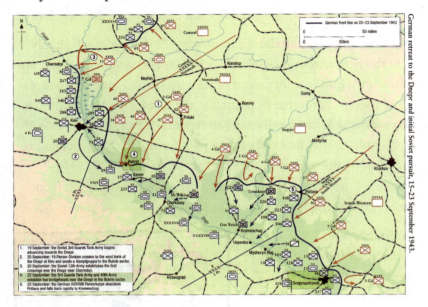

German retreat to the Dnepr and initial Soviet pursuit, 15–23 September 1943.

The Soviets managed to get more than 100,000 men across the Dnieper River, but there was intense fighting, and at the end of September the 206th Rifle Division (as part of the 47th Army) was probably involved in the later stages of the fight for the Bukrin Bridgehead. And this is probably where Gregory Schönfinkel was wounded.

After being wounded, he seems to have been taken to some kind of service area for the 206th Rifle Division (206 зсп ЗапФ), from which he was picked up by a hospital train (and, yes, it was actually a moving hospital, with lots of cars with red crosses painted on top).

But more significant in our quest for the story of Gregory Schönfinkel is other information in the military documents we have. They record that he is Jewish (as opposed to "Russian", which is how basically all the other soldiers in these lists are described). Then they say that he has "higher education". One says he is an "engineer". Another is more specific, and says he's an "engineer economist" (Инж. Эконом.). They also say that he is not a member of the Communist Party.

They say he is a widower, and that his wife's name was Evdokiya Ivanovna (Евдокия Иван.). They also list his "mother", giving her name as Мария Григ. ("Maria Grig.", perhaps short for "Grigorievna"). And then they list an address: Москва С. Набер. д. 26 кв. 1ч6, which is presumably 26 Sofiyskaya Embankment, Apartment 1-6, Moscow.

Where is that address? Well, it turns out it's in the very center of Moscow ("inside the Garden Ring"), with the front looking over the Moscow River directly at the Kremlin:

Here's a current picture of the building

as well as one from perhaps 100 years earlier:

The building was built by a family of merchants named the Bakhrushins in 1900–1903 to provide free apartments for widows and orphans (apparently there were about 450 one-room 150-to-300-square-foot apartments). In the Russian Revolution, the building was taken over by the government, and set up to house the Ministry of Oil and Gas. But some "communal apartments" were left, and it's presumably in one of those that Gregory Schönfinkel lived. (Today the building is the headquarters of the Russian state oil company Rosneft.)

OK, but let's unpack this a bit further. "Communal apartments" basically means dormitory-style housing. A swank building, but apparently not so swank accommodation. Well, actually, in Soviet times dormitory-style housing was pretty typical in Moscow, so this really was a swank setup.

But then there are a couple of mysteries. First, how come a highly educated engineering economist with a swank address was just a private in the army? (When the hospital train picked up Gregory, along with four other privates, one of the others was listed as a carpenter; the others were all listed as "с/хоз" or "сельское хозяйство", basically meaning "farm laborer", or what before Soviet times would have been called "peasant").

Maybe the Russian army was so desperate for recruits after all their losses that—despite being 44 years old—Gregory was drafted. Maybe he volunteered (though then we have to explain why he didn't do that earlier). But regardless of how he wound up in the army, maybe his status as a private had to do with the fact that he wasn't a member of the Communist Party. At that time, a large fraction of the city-dwelling "elite" were members of the Communist Party (and it wouldn't have been a major problem that he was Jewish, though coming from a merchant family might have been a negative). But if he wasn't in the "elite", how come the swank address?

A first observation is that his wife's first name *Evdokiya* was a popular Russian Orthodox name, at least before 1917 (and is apparently popular again now). So presumably Gregory had—not uncommonly in the Soviet era—married someone who wasn't Jewish. But now let's look at the "mother's" name: "Мария Григ." ("Maria Grig.").

We know Gregory's (and Moses's) mother's name was Maria/"Masha" Gertsovna Schönfinkel (née Lurie)—or Мария ("Маша") Герцовна Шейнфинкель. And according to other information, she died in 1936. So—unless someone miswrote Gregory's "mother's" name—the patronymics (second names) don't match. So what's going on?

My guess is that the "mother" is actually a mother-in-law, and that it was her apartment. Perhaps her husband (most likely at that point not her) had worked at the Ministry of Oil and Gas, and that's how she ended up with the apartment. Maybe Gregory worked there too.

OK, so what was an "engineer economist" (Инженер Экономист)? In the planning-oriented Soviet system, it was something quite important: basically a person who planned and organized production and labor in some particular industry.

How did one become an "engineer economist"? At least a bit later, it was a 5-year "master's level" course of study, including courses in engineering, mathematics, bookkeeping, finance, economics of a particular sector, and "political economy" (à la Marx). And it was a very Soviet kind of thing. So the fact that that was what Gregory did presumably means that he was educated in the Soviet Union.

He must have finished high school right when the Tsar was being overthrown. Probably too late to be involved in World War I. But perhaps he got swept up in the Russian Civil War. Or maybe he was in college then, getting an early Soviet education. But, in any case, as an engineer economist it's pretty surprising that in World War II he didn't get assigned to something technical in the army, and was just a simple private in the infantry.

From the data we have, it's not clear what was going on. But maybe it had something to do with Moses.

It's claimed that Moses died in 1940 or 1942 and was "living in a communal apartment". Well, maybe that communal apartment was actually Gregory's (or at least his mother-in-law's) apartment. And here's a perhaps fanciful theory: Gregory joined the army out of some kind of despondency. His wife died. His older brother died. And in February 1942 (though it might have taken him a while to find out) any of his family still in Ekaterinoslav probably died in the massacre of the Jewish population there (at least if they hadn't evacuated as a result of earlier bombing). He hadn't joined the army earlier in the war, notably during the Battle of Moscow. And by 1943 he was 44 years old. So perhaps in some despondency—or anger—he volunteered for the army.

We don't know. And at this point the trail seems to go cold. It doesn't appear that Gregory had any children, and we haven't been able to find out anything more about him.

But I consider it progress that we've managed to identify that Moses's younger brother lived in Moscow, potentially providing a plausible reason that Moses might have gone to Moscow.

Actually, there may have been other "family reasons". There seems to have been quite a lot of back-and-forth in the Jewish population between Moscow and Ekaterinoslav. And Moses's mother came from the Lurie family, which was prominent not only in Ekaterinoslav, but also in Moscow. And it turns out that the Lurie family has done a fair amount of genealogy research. So we were able, for example, to reach a first cousin once removed of Moses's (i.e. someone whose parent shared a grandparent with Moses, or 1/32 of the genetics). But so far nobody has known anything about what happened to Moses, and nobody has said "Oh, and by the way, we have a suitcase full of strange papers" or anything.

I haven't given up. And I'm hoping that we'll still be able to find out more. But this is where we've got so far.

One More Thing

In addition to pursuing the question of the fate of Moses Schönfinkel, I've made one other potential connection. Partly in compiling a bibliography of combinators, I discovered a whole collection of literature about "combinatory categorial grammars" and "combinatory linguistics".

What are these? These days, the most common way to parse an English sentence like "I am trying to track down a piece of history" is a hierarchical tree structure—analogous to the way a context-free computer language would be parsed:

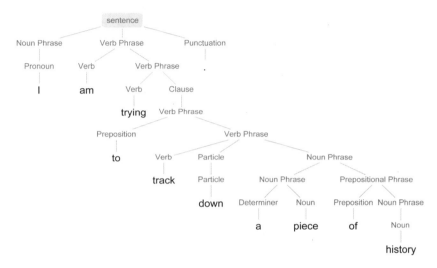

But there is an alternative—and, as it turns out, significantly older—approach: to use a so-called dependency grammar in which verbs act like functions, "depending" on a collection of arguments:

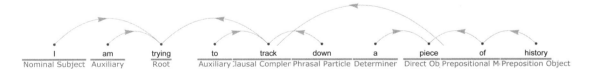

In something like Wolfram Language, the arguments in a function would appear in some definite order and structure, say as $f[x, y, z]$. But in a natural language like English, everything is just given in sequence, and a function somehow has to have a way to figure out what to grab. And the idea is that this process might work like how combinators written out in sequence "grab" certain elements to act on.

This idea seems to have a fairly tortuous history, mixed up with attempts and confusions about connecting the syntax (i.e. grammatical structure) of human languages to their semantics (i.e. meaning). The core issue has been that it's perfectly possible to have a syntactically correct sentence ("The flying chair ate a happy semicolon") that just doesn't seem to have any "real-world" meaning. How should one think about this?

I think the concept of computational language that I've spent so many years developing actually makes it fairly clear. If one can express something in computational language there's a way to compute from it. Maybe the resulting computation will align with what happens in the real world; maybe it won't. But there's some "meaningful place to go" with what one has. And the point is that a computational language has a well-defined "inner computational representation" for things. The particular syntax (e.g. sequence of characters) that one might use for input or output in the computational language is just something superficial.

But without the idea of computational language people have struggled to formalize semantics, tending to try to hang what they're doing on the detailed structure and syntax of human languages. But then what should one do about syntactically correct structures that don't "mean anything"? An example of what I consider to be a rather bizarre solution—embodied in so-called Montague grammars from the 1970s—is essentially to turn pieces of certain sentences into functions, in which there's nothing "concrete" there, just "slots" where things could go ("$x\_$ ate $y\_$")—and where one can "hold off meaninglessness" by studying things without explicitly filling in the slots.

In the original formulation, the "functions" were thought about in terms of lambdas. But combinatory categorial grammars view them instead in terms of combinators, in which in the course of a sentence words in a sense "apply to each other". And even without the notion of slots one can do "combinatory linguistics" and imagine finding the structure of sentences by taking words to "apply themselves" "across the sentence" like combinators.

If well designed (as I hope the Wolfram Language is!) computational language has a certain clean, formal structure. But human natural language is full of messiness, which has to be untangled by natural language understanding—as we've done for so many years for Wolfram|Alpha, always ultimately translating to our computational language, the Wolfram Language.

But without the notion of an underlying computational language, people tend to feel the need to search endlessly for formal structure in human natural language. And, yes, some exists. But—as we see all the time in actually doing practical natural language understanding for Wolfram|Alpha—there's a giant tail that seems to utterly explode any all-encompassing formal theory.

Are there at least fragments that have formal structure? There are things like logic ("and", "or", etc.) that get used in human language, and which are fairly straightforwardly formalizable. But maybe there are more "functional" structures too, perhaps having to do with the operation of verbs. And in combinatory linguistics, there've been attempts to find these—even for example directly using things like Schönfinkel's S combinator. (Given $S f g x \to f[x][g[x]]$ one can start imagining—with a slight stretch—that "eat peel orange" operates like the S combinator in meaning "eat[orange][peel[orange]]".)

Much of the work on this has been done in the last few decades. But it turns out that its history stretches back much further, and might conceivably actually intersect with Moses Schönfinkel himself.

The key potential link is Kazimierz Ajdukiewicz (1890–1963). Ajdukiewicz was a Polish logician/philosopher who long tried to develop a "mathematicized theory" of how meaning emerges, among other things, from natural language, and who basically laid the early groundwork for what's now combinatory linguistics.

Kazimierz Ajdukiewicz was born two years after Moses Schönfinkel, and studied philosophy, mathematics and physics at the University of Lviv (now in Ukraine), finishing his PhD in 1912 with a thesis on Kant's philosophy of space. But what's most interesting for our purposes is that in 1913 Ajdukiewicz went to Göttingen to study with David Hilbert and Edmund Husserl.

In 1914 Ajdukiewicz published one paper on "Hilbert's New Axiom System for Arithmetic", and another on contradiction in the light of Bertrand Russell's work. And then in 1915 Ajdukiewicz was drafted into the Austrian army, where he remained until 1920, after which he went to work at the University of Warsaw.

But in 1914 there's an interesting potential intersection. Because June of that year is when Moses Schönfinkel arrived in Göttingen to work with Hilbert. At the time, Hilbert was mostly lecturing about physics (though he also did some lectures about "principles of mathematics"). And it seems inconceivable that—given their similar interests in the structural foundations of mathematics—they wouldn't have interacted.

Of course, we don't know how close to combinators Schönfinkel was in 1914; after all, his lecture introducing them was six years later. But it's interesting to at least imagine some interaction with Ajdukiewicz. Ajdukiewicz's own work was at first most concerned with things like the relationship of mathematical formalism and meaning. (Do mathematical constructs "actually exist", given that their axioms can be changed, etc.?) But by the beginning of the 1930s he was solidly concerned with natural language, and was soon writing papers with titles like "Syntactic Connexion" that gave formal symbolic descriptions of language (complete with "functors", etc.) quite reminiscent of Schönfinkel's work.

So far as I can tell Ajdukiewicz never explicitly mentioned Schönfinkel in his publications. But it seems like too much of a coincidence for the idea of something like combinators to have arisen completely independently in two people who presumably knew each other—and never to have independently arisen anywhere else.

Thanks

Thanks to Vitaliy Kaurov for finding additional documents (and to the State Archives of the Dnipropetrovsk Region and Elena Zavoiskaia for providing various documents), Oleg and Anna Marichev for interpreting documents, and Jason Cawley for information about military history. Thanks also to Oleg Kiselyov for some additional suggestions on the original version of this piece.

THE WOLFRAM
S COMBINATOR CHALLENGE

$20,000
PRIZE

Is the S combinator computation universal?

Prize Announcement by Stephen Wolfram »

$$S\,\varphi\,\chi\,x = (\varphi\,x)(\chi\,x)$$

e S, K combinators defined by Moses Schönfinkel on December 7, 1920
ogether known to be computation universal. On December 7, 2020
en Wolfram made the suggestion that S alone might also be universal.
allenge is about proving or disproving that conjecture.

oduction to Combinators »
d the Story of Computation »

s Schönfinkel »

1920, 2020 and a $20,000 Prize: Announcing the S Combinator Challenge

Hiding in Plain Sight for a Century?

On December 7, 1920, Moses Schönfinkel introduced the S and K combinators— and in doing so provided the first explicit example of a system capable of what we now call universal computation. A hundred years later—as I prepared to celebrate the centenary of combinators—I decided it was time to try using modern computational methods to see what we could now learn about combinators. And in doing this, I got a surprise.

It's already remarkable that S and K yield universal computation. But from my explorations I began to think that something even more remarkable might be true, and that in fact S alone might be sufficient to achieve universal computation. Or in other words, that just applying the rule

$$S f g x \rightarrow f[x][g[x]]$$

over and over again might be all that's needed to do any computation that can be done.

I don't know for sure that this is true, though I've amassed empirical evidence that seems to point in this direction. And today I'm announcing a prize of $20,000 (yes, the "20" goes with the 1920 invention of combinators, and the 2020 making of my conjecture) for proving—or disproving—that the S combinator alone can support universal computation.

Why is it important to know? Obviously it'll be neat if it turns out that hiding in plain sight for a century has been an even simpler basis for universal computation than we ever knew. But more than that, determining whether the S combinator alone is universal will provide an important additional data point in the effort to map out just where the threshold of universal computation lies.

Practical computers have complicated and carefully designed CPUs. But is that complexity really needed to support universal computation? My explorations in the computational universe over the past 40 years have led me to the conclusion that it absolutely is not, and that in fact even among systems with the very simplest rules, universal computation is actually quite ubiquitous.

I've developed a very general principle that I call the Principle of Computational Equivalence that implies that pretty much whenever we see behavior that isn't in some sense obviously simple, then it will actually correspond to a computation that's equivalent in sophistication to any other computation. And one of the many consequences of this principle is that it says computation universality should be ubiquitous.

For more than 30 years I've been pushing to get explicit evidence about this, by proving (or disproving) that particular systems are computation universal. And from this have come two very excellent examples that help validate the Principle of Computational Equivalence: the rule 110 cellular automaton

and the 2, 3 Turing machine:

Both these systems I identified as the simplest of their type that could conceivably be computation universal—and both I expected would actually be computation universal.

But in both cases it was hard work to come up with an explicit proof. In the first case, the proof was part of my book *A New Kind of Science* (with much of the heavy lifting done by a then-research assistant of mine). In the second case, I'd already "made the conjecture" in *A New Kind of Science*, but in 2007 I decided to put up a prize for actually resolving it. And I was very pleased when, after just a few months, a proof was found, and the prize was won.

So now I'm hoping something similar will happen in the S combinator case. I'm expecting that it will turn out to be computation universal, but it would perhaps be even more interesting if it was proved not to be.

In the past one might have viewed proving that a simple system is (or is not) computation universal as a curiosity—something like solving a specific mathematical puzzle. But the Principle of Computational Equivalence—and all the science I've done as a result of exploring the computational universe—now places such a question in a broader context, and shows that far from being a curiosity, it's actually quite central.

For example, with computation universality comes computational irreducibility, and undecidability. And to know how common these are in systems in nature, in mathematics and in technology has great fundamental and practical significance for when predictions can be made, and what kinds of questions can be answered.

Identifying the threshold of computation universality also has direct implications for making a molecular-scale computer, as well, for example, as in computer security, for knowing whether some part of some program is actually universal, so it can be programmed to do something bad.

The Basic Setup

In the usual S, K combinator setup one does computations by setting up an initial combinator expression that encodes the computation one wants to do, then one repeatedly applies the combinator rules until one gets to a fixed point that can be interpreted as the result of the computation. Of course, not all computations halt, and for computations that don't halt, no fixed point will ever be reached.

With the S combinator alone a direct input → output representation of computations won't work. The reason is that there is an elaborate, but finite, characterization of every S combinator expression that halts, and the existence of this characterization implies that halting S combinator evolutions don't have enough richness to represent arbitrary computations.

There are, however, plenty of non-halting S combinator evolutions. Here's the very simplest of them:

SSS(SS)SS

S(SS)(S(SS))SS

SSS(S(SS)S)S

S(S(SS)S)(S(S(SS)S))S

S(SS)SS(S(S(SS)S)S)

SSS(SS)(S(S(SS)S)S)

S(SS)(S(SS))(S(S(SS)S)S)

SS(S(S(SS)S)S)(S(SS)(S(S(SS)S)S))

S(S(SS)(S(S(SS)S)S))(S(S(SS)S)S(S(SS)(S(S(SS)S)S)))

And here are examples of the sequences of sizes of expressions obtained in various such evolutions:

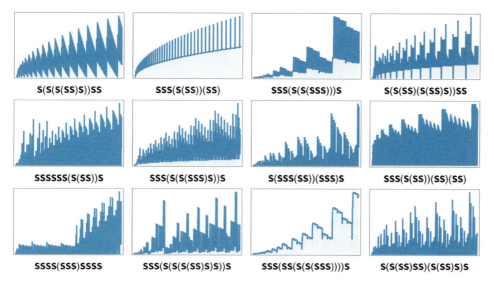

S(S(S(SS)S))SS	SSS(S(SS))(SS)	SSS(S(S(SSS)))S	S(S(SS)(S(SS)S))SS
SSSSSS(S(SS))S	SSS(S(S(SSS)S))S	S(SSS(SS))(SSS)S	SSS(S(SS))(SS)(SS)
SSSS(SSS)SSSS	SSS(S(S(S(SS)S)S))S	SSS(SS(S(S(SSS))))S	S(S(SS)SS)(S(SS)S)S

And potentially it's possible to do computations by "piggy-backing" on these non-halting evolutions. Start with a specification of the computation you want to do. Now encode it as an initial condition for one of the infinite number of possible non-halting combinator evolutions. Then run the combinator evolution until some particular feature occurs. Then decode from the results at that stage the output from the computation.

To prove that the S combinator is capable of supporting universal computation, one must show that something like this will work for any possible computation one wants to do. Or, in other words, one must show that an S combinator system can emulate some other system (say the S, K combinator one) that is already known to be universal. And by "emulate", what we mean here is that a suitable "encoder", "detector" and "decoder" can be found that will allow any evolution in the system being emulated to be mapped to a corresponding evolution in the S combinator system.

There is an important caveat, however: one has to be sure that the encoder, detector and decoder are not themselves capable of universal computation. If one's lucky, these will operate in sufficiently straightforward ways so that it'll be obvious they're not universal. But in general it'll be nontrivial to demonstrate that they're not universal. Depending on the setup, one approach may be to show that for all inputs they must halt in some bounded time.

OK, so what if the S combinator system is not computation universal? How could one show that? One possibility would be to demonstrate that all, even infinite, evolutions have only a finite, or essentially finite, number of distinct forms. Another would be to establish that any "modulation" of such evolution must always die out in finite time, so that it must in effect correspond to a computation that halts.

A slightly different approach would be to show that S combinator evolution can be "solved" to the point where its properties after any number of steps can be determined by some bounded computation.

We've been talking about "doing computations" with combinators. But there's a subtlety with that. Because there are in general many different possible evaluation orders for the combinator evolution. If the evolution halts, then it's guaranteed that it'll always reach the same fixed point, regardless of the order in which the evaluation was done. But if it doesn't halt, there's in general a whole multiway graph of possible sequences of results.

So what does computation universality then mean? As I've discussed elsewhere, a strong potential definition is to ask that the multiway system generated by the combinator system can effectively emulate the multiway system generated, say, by a multiway (or nondeterministic) Turing machine. But a weaker alternative definition might just be to ask whether one can find some path (i.e. some evaluation order) that can perform any given (deterministic) computation. Or one might for example have detector and decoder functions that can work on some—or all—branches, somehow always being able to reproduce the computation one wants.

The Operation of the S Combinator Challenge

I don't know how long it's going to take, or how hard it's going to be, but I'm hoping that someday a message will arrive that successfully resolves the S Combinator Challenge. What will the message have to contain?

The best case as far as I am concerned is specific Wolfram Language code that implements the solution. If the S combinator is in fact universal, then the code should provide a "compiler" that takes, say, a Turing machine or cellular automaton specification, and generates an S combinator initial condition, together with code for the "detector" and "decoder". But how will we validate that this code is "correct"?

It might conceivably be simple enough that it's amenable to direct human analysis. But more likely it's going to require some level of automated analysis. At first one might just do explicit testing. But then one might for example try to generate an automated proof that the detector and decoder always halt, and even perhaps determine an upper bound on how long this can take.

If the S combinator is not universal, one might show this by having code that can "jump ahead" by any number of steps and explicitly determine the outcome, but that itself always halts. Or maybe one might have code that effectively implements what happens to any "modulation" of S combinator evolution—and then prove that this code leads to behavior that, say, always halts, or otherwise ultimately shows complete regularity.

But maybe one won't have explicit code that implements a solution, and instead one will only have an abstract proof of what's possible. This could conceivably be in the form of a standard human-oriented mathematical proof. But I think it's more likely it will be complicated enough that it has to be presented in a systematic computational way. And in the end, if it's a proof it'll have to start from certain axioms.

What axioms can be used? One would hope that the "standard axioms of mathematics" will be enough. Perhaps only the Peano axioms will be needed. But I wouldn't be surprised if transfinite induction were needed, requiring the axioms of set theory. And of course there are weird cases that could arise, where the proof will be possible, say, with one assumption about the Continuum Hypothesis, but not with the other.

There are other weird things that could happen. For example, it might be possible to show that the complete infinite structure obtained by running the combinator evolution for an infinite time could reproduce any computation, but one might not know whether this is possible with finite evolution. Or it might be that one could only make a construction if one had an infinite tree (say specified by a transfinite number) as the initial condition.

It's also conceivable that the behavior of the S combinator could correspond to an "intermediate degree" of computational capability—and have undecidable features, but not be universal. I don't think this will happen, but if it does, I would consider it a resolution of the S Combinator Challenge.

And then, of course, there's the unexpected. And in my experience of exploring the computational universe over the past 40 years, that's something one encounters with remarkable frequency. A corner case one didn't anticipate. An intermediate situation that didn't seem possible. A new form of behavior one never imagined. Sometimes these are the most interesting things to find. But inevitably they make giving a cut-and-dried definition of what's needed for the S Combinator Challenge difficult.

It's worth remembering, though, that the S Combinator Challenge is about answering the human-stated question "Is the S combinator computation universal?". And in the end it may take humans to determine whether some particular potential solution really does answer that question or not.

In a sense the S Combinator Challenge is already a 100-year story. But I'm certainly hoping that it'll be resolved in far less than 100 more years. And that in doing so we'll learn important things about the remarkable world of combinators, and the computational universe in general.

www.combinatorprize.org

STEPHEN WOLFRAM
A NEW KIND OF SCIENCE

Excerpts from
A New Kind of Science
(2002)

CHAPTER 3: The World of Simple Programs
Section 10: Symbolic Systems

Register machines provide simple idealizations of typical low-level computer languages. But what about [Wolfram Language]? How can one set up a simple idealization of the transformations on symbolic expressions that [Wolfram Language] does? One approach suggested by the idea of combinators from the 1920s is to consider expressions with forms such as $e[e[e][e]][e]\ e$ and then to make transformations on these by repeatedly applying rules such as $e[x\_][y\_] \rightarrow x[x\_][y\_]]$, where $x\_$ and $y\_$ stand for any expression.

The picture below shows an example of this. At each step the transformation is done by scanning once from left to right, and applying the rule wherever possible without overlapping.

A sequence of steps in the evolution of a simple symbolic system. At each step each boxed region is transformed according to the rule shown. This transformation corresponds to applying the basic [Wolfram Language] operation expression /. rule

The structure of expressions like those on the facing page is determined just by their sequence of opening and closing brackets. And representing these brackets by dark and light squares respectively, the picture below shows the overall pattern of behavior generated.

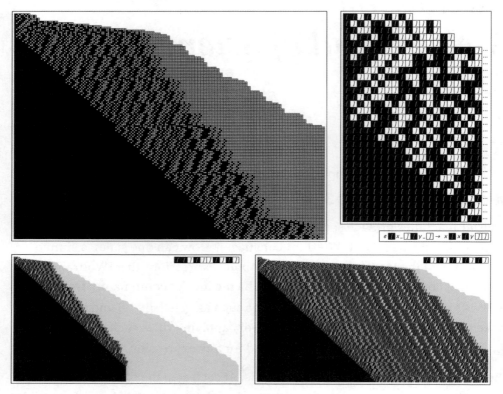

More steps in the evolution on the previous page, with opening brackets represented by dark squares and closing brackets by light ones. In each case configurations wider than the picture are cut off on the right. For the initial condition from the previous page, the system evolves after 264 steps to a fixed configuration involving 256 opening brackets followed by 256 closing brackets. For the initial condition on the bottom right, the system again evolves to a fixed configuration, but now this takes 65,555 steps, and the configuration involves 65,536 opening and closing brackets. Note that the evolution rules are highly non-local, and are rather unlike those, say, in a cellular automaton. It turns out that this particular system always evolves to a fixed configuration, but for initial conditions of size n can take roughly n iterated powers of 2 (or $2^{2^{2}}$) to do so.

With the particular rule shown, the behavior always eventually stabilizes—though sometimes only after an astronomically long time.

But it is quite possible to find symbolic systems where this does not happen, as illustrated in the pictures below. Sometimes the behavior that is generated in such systems has a simple repetitive or nested form. But often—just as in so many other kinds of systems—the behavior is instead complex and seemingly quite random.

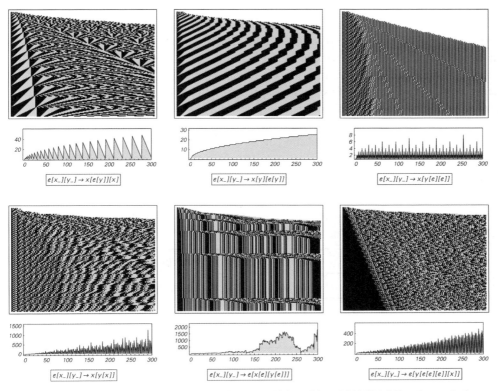

The behavior of various symbolic systems starting from the initial condition *e[e[e][e]][e] e]*. The plots at the bottom show the difference in size of the expressions obtained on successive steps.

NOTES

Implementation of symbolic systems.

The evolution for *t* steps of the first symbolic system shown can be implemented simply by

NestList[#/.e[x_][y_]→x[x[y]]&,init,t]

Symbolic expressions. Expressions like *Log[x]* and *f[x]* that give values of functions are familiar from mathematics and from typical computer languages. Expressions like *f[g[x]]* giving compositions of functions are also familiar. But in general, as in [Wolfram Language], it is possible to have expressions in which the head *h* in *h[x]* can itself be any expression—not just a single symbol. Thus for example *f[g[x]]*, *f[g[h]][x]* and *f[g][h][x]* are all possible expressions. And these kinds of expressions often arise in [Wolfram Language] when one manipulates functions as a whole before applying them to arguments. ($\partial_{xx} f[x]$ for example gives *f"[x]* which is *Derivative[2][f][x]*.) (In principle one can imagine representing all objects with forms such as *f[x, y]* by so-called currying as *f[x][y]*, and indeed I tried this in the early 1980s in SMP. But although this can be convenient when *f* is a discrete function such as a matrix, it is inconsistent with general mathematical and other usage in which for example *Gamma[x]* and *Gamma[a, x]* are both treated as values of functions.)

Representations for symbolic expressions. Among the representations that can be used for expressions are:

functional	a[b[c[d]]]	a[b][c[d]]	a[b[c]][d]	a[b][c][d]
Polish	{∘, a, ∘, b, ∘, c, d}	{∘, ∘, a, b, ∘, c, d}	{∘, a, ∘, ∘, b, c, d}	{∘, ∘, ∘, a, b, c, d}
operator	a∘(b∘(c∘d))	(a∘b)∘(c∘d)	a∘((b∘c)∘d)	((a∘b)∘c)∘d
tree	{a, {b, {c, d}}}	{{a, b}, {c, d}}	{a, {{b, c}, d}}	{{{a, b}, c}, d}

Typical transformation rules are non-local in all these representations. Polish representation (whose reverse form has been used in HP calculators) for an expression can be obtained using

Flatten[expr //. x_[y_] → {∘, x, y}]

The original expression can be recovered using

First[Reverse[list] //. {w___, x_, y_, ∘, z___} → {w,y[x],z}]

(Pictures of symbolic system evolution made with Polish notation differ in detail but look qualitatively similar to those made as in the main text with functional notation.)

The tree representation of an expression can be obtained using *expr //. x_[y_] → {x, y}*, and when each object has just one argument, the tree is binary, as in LISP.

If only a single symbol ever appears, then all that matters is the overall structure of an expression, which can be captured as in the main text by the sequence of opening and closing brackets, given by

Flatten[Characters[ToString[expr]] /. {"[" → 1, "]" → 0, "e" → {}}]

Enumerating possible expressions. *LeafCount[expr]* gives the number of symbols that appear anywhere in an expression, while *Depth[expr]* gives the number of closing brackets at the end of its functional representation—equal to the number of levels in the rightmost branch of the tree representation. (The maximum number of levels in the tree can be computed from

expr /. _Symbol → 1 //. x_[y_] → 1 + Max[x, y].)

With a list *s* of possible symbols, *c[s, n]* gives all possible expressions with *LeafCount[expr] == n*:

c[s_, 1] = s; c[s_, n_] := Flatten[
Table[Outer[#1[#2] &, c[s, n -m], c[s, m]], {m, n - 1}]]

There are a total of such *Binomial[2 n - 2, n - 1] Length[s]^n / n* expressions. When *Length[s] == 1* the expressions correspond to possible balanced sequences of opening and closing brackets (see page 989 [in *A New Kind of Science*]).

Properties of example symbolic system. All initial conditions eventually evolve to expressions of the form *Nest[e, e, m]*, which then remain fixed. The quantity *expr //. {e → 0, x_[y_] → 2 x + y}* turns out to remain constant through the evolution, so this gives the final value of *m* for any initial condition. The maximum is *Nest[2$^\#$ &, 0, n]*, achieved for initial conditions of the form *Nest[#[e] &, e, n]*. (By analogy with page [320] any expression can be interpreted as a Church numeral *u = expr //. {e → 2, x_[y_] → y^x} = 2$^{2^m}$*, so that *expr[a][b]* evolves to *Nest[a, b, u]*.) During the evolution the rule can apply only to the inner part *FixedPoint[Replace[#, e[x_] → x] &, expr]* of an expression. The depth of this inner part for initial condition *e[e][e][e][e][e]* is shown below. For all initial conditions this depth seems at first to increase linearly, then to decrease in a nested way according to

FoldList[Plus, 0, Flatten[Table[
{1, 1, Table[-1, {IntegerExponent[i, 2] +1}]}, {i, m}]]]

This quantity alternates between value *1* at position *2^i* and value *j* at position *2^i - j + 1*. It reaches a fixed point as soon as the depth reaches 0. For initial conditions of size *n*, this occurs after at most *Sum[Nest[2$^\#$ &, 0, i] - 1, {i, n}] + 1* steps.

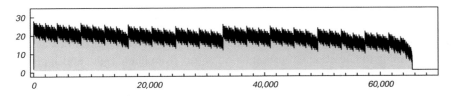

Other symbolic systems rules. If only a single variable appears in the rule, then typically only nested behavior can be generated—though in an example like *e[x_][_] → e[x[e[e]][e]][e]]* it can be quite complex. The left-hand side of each rule can consist of any expression; *e[e[x_]][y_]* and *e[e][x_[y_]]* are two possibilities. However, at least with small initial conditions it seems easier to achieve complex behavior with rules based on *e[x_][y_]*. Note that rules with no explicit *e*'s

on the lefthand side always give trees with regular nested structures; *x_[y_] → x[y][x[y]]* (or
x_ → x[x] in [Wolfram Language]), for example, yields balanced binary trees.

Long halting times in symbolic systems. Symbolic systems with rules of the form
e[x_][y_] → Nest[x, y, r] always evolve to fixed points—though with initial conditions of size *n*
this can take of order *Nest[r# &, 0, n]* steps (see above). In general there will be symbolic
systems where the number of steps to evolve to a fixed point grows arbitrarily rapidly with *n*
(see page 1145 [in *A New Kind of Science*]), and indeed I suspect that there are even systems
with quite simple rules where proving that a fixed point is always reached in a finite number
of steps is beyond, for example, the axiom system for arithmetic (see page 1163 [in *A New
Kind of Science*]).

Trees representation for symbolic systems. The rules given on pages [308 and 309]
correspond to the transformations on trees shown below.

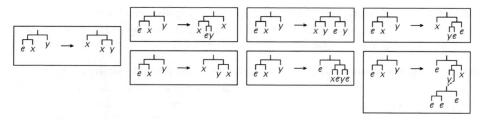

The first few steps in evolution from two initial conditions of the system on page [308]
correspond to the sequences of trees below.

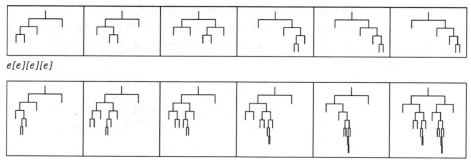

e[e][e][e]

e[e[e][e]][e][e]

Order dependence in symbolic systems. The operation *expr /. lhs → rhs* in [Wolfram Language]
has the effect of scanning the functional representation of *expr* from left to right, and applying
rules whenever possible while avoiding overlaps. (Standard evaluation in [Wolfram Language]
is equivalent to *expr //. rules* and uses the same ordering, while *Map* uses a different order.) One
can have a rule be applied only once using

   Module[{i = 1}, expr /. lhs :> rhs /; i++==1]

Many symbolic systems (including the one on page [308]) have the so-called Church–Rosser
property (see page [57, etc.]) which implies that if a fixed point is reached in the evolution of
the system, this fixed point will be the same regardless of the order in which rules are applied.

History of symbolic systems. Symbolic systems of the general type I discuss here seem to have first arisen in 1920 in the work of Moses Schönfinkel on what became known as combinators. As discussed on page [319] Schönfinkel introduced certain specific rules that he suggested could be used to build up functions defined in logic. Beginning in the 1930s there were a variety of theoretical studies of how logic and mathematics could be set up with combinators, notably by Haskell Curry. For the most part, however, only Schönfinkel's specific rules were ever used, and only rather specific forms of behavior were investigated. In the 1970s and 1980s there was interest in using combinators as a basis for compilation of functional programming languages, but only fairly specific situations of immediate practical relevance were considered. (Combinators have also been used as logic recreations, notably by Raymond Smullyan.)

Constructs like combinators appear to have almost never been studied in mainstream pure mathematics. Most likely the reason is that building up functions on the basis of the structure of symbolic expressions has never seemed to have much obvious correspondence to the traditional mathematical view of functions as mappings. And in fact even in mathematical logic, combinators have usually not been considered mainstream. Most likely the reason is that ever since the work of Bertrand Russell in the early 1900s it has generally been assumed that it is desirable to distinguish a hierarchy of different types of functions and objects— analogous to the different types of data supported in most programming languages. But combinators are set up not to have any restrictions associated with types. And it turns out that among programming languages [Wolfram Language] is almost unique in also having this same feature. And from experience with [Wolfram Language] it is now clear that having a symbolic system which—like combinators—has no built-in notion of types allows great generality and flexibility. (One can always set up the analog of types by having rules only for expressions whose heads have particular structures.)

Operator systems. One can generalize symbolic systems by having rules that define transformations for any [Wolfram Language] pattern. Often these can be thought of as one-way versions of axioms for operator systems (see page 1171 [in *A New Kind of Science*]), but applied only once per step (as /. does), rather than in all possible ways (as in a multiway system)—so that the evolution is just given by *NestList[# /. rule &, init, t]*. The rule $x\_ \to x \circ x$ then for example generates a balanced binary tree. The pictures below show the patterns of opening and closing parentheses obtained from operator system evolution rules in a few cases.

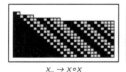 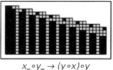 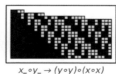 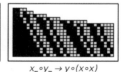

$x\_ \to x \circ x$ $x\_ \circ y\_ \to (y \circ x) \circ y$ $x\_ \circ y\_ \to (y \circ y) \circ (x \circ x)$ $x\_ \circ y\_ \to y \circ (x \circ x)$

Network analogs of symbolic systems. The state of a symbolic system can always be viewed as corresponding to a tree. If a more general network is allowed then rules based on analogs of network substitution systems from from page 508 [in *A New Kind of Science*] can be used. (One can also construct an infinite tree from a general network by following all its possible paths, but in most cases there will be no simple way to apply symbolic system rules to such a tree.)

Excerpt from **Section 12: Universality in Turing Machines and Other Systems**

⋮

I suspect that in almost any case where we have seen complex behavior earlier in this book it will eventually be possible to show that there is universality. And indeed, as I will discuss at length in the next chapter, I believe that in general there is a close connection between universality and the appearance of complex behavior.

Previous examples of systems that are known to be universal have typically had rules that are far too complicated to see this with any clarity. But an almost unique instance where it could potentially have been seen even long ago are what are known as combinators.

Combinators are a particular case of the symbolic systems that we discussed on page [307]. Originally intended as an idealized way to represent structures of functions defined in logic, combinators were actually first introduced in 1920— sixteen years before Turing machines. But although they have been investigated somewhat over the past eighty years, they have for the most part been viewed as rather obscure and irrelevant constructs.

The basic rules for combinators are given below.

$$s[x\_][y\_][z\_] \to x[z][y[z]]$$
$$k[x\_][y\_] \to x$$

Rules for symbolic systems known as combinators, first introduced in 1920, and proved universal by the mid-1930s.

With short initial conditions, the pictures at the top of the next page demonstrate that combinators tend to evolve quickly to simple fixed points. But with initial condition (e) of length 8 the pictures show that no fixed point is reached, and instead there is exponential growth in total size—with apparently rather random internal behavior.

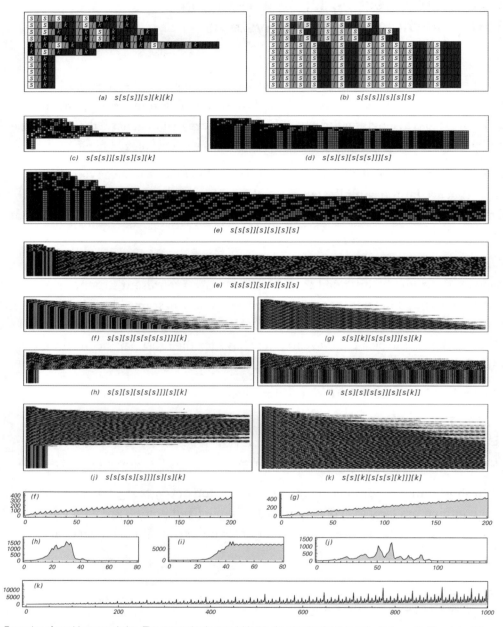

Examples of combinator evolution. The expression in case (e) is the shortest that leads to unlimited growth. The plots at the bottom show the total sizes of expressions reached on successive steps. Note that the detailed pattern of evolution—though not any final fixed point reached—can depend on the fact that the combinator rules are applied at each step in [Wolfram Language] /. order.

Other combinators yield still more complicated behavior—sometimes with overall repetition or nesting, but often not.

There are features of combinators that are not easy to capture directly in pictures. But from pictures like the ones on the facing page it is rather clear that despite their fairly simple underlying rules, the behavior of combinators can be highly complex.

And while issues of typical behavior have not really been studied before, it has been known that combinators are universal almost since the concept of universality was first introduced in the 1930s.

One way that we can now show this is to demonstrate that combinators can emulate rule 110. And as the pictures on the next page illustrate, it turns out that just repeatedly applying the combinator expression below reproduces successive steps in the evolution of rule 110.

```
s[s[k[s]]][s[k[s[s[k][k]]]]][s[k[k]]][s[s[s[s[s[k][k]][k[s[k]]]]][k[s[s[k[s]]][s[k[s[s[k][k]]]]][s[k[k]]][s[s[k[
s]]][s[k[s[s[k][k]]]]][s[k[k]]][s[s[k][k]][k[k]]]]]]][s[k[k]]][s[s[s[k][k]][k[s[k]]]][k[s[k]]]]]]]]]][s[k[k]]][s[s[
s[k][k]][k[s[s[s[s[k][k]]][k[s[s[s[s[k][k]][k[s[s[k][k]][k[s[s[s[k][k]]]]][s[k][k]][s[s[s[k][k]][k[k]]][k[k]]]]]]][s[k[
k]]][s[s[s[s[k][k]][k[s[k]]]][k[s[s[k][s]]][s[k[s[s[k][k]]]]][s[k[k]]][s[s[k][k]][k[k]]]][k[k]]]]]]][s[k[k]]][
s[s[k[s]]][s[k[s[s[k][k]]]]][s[k[k]]][s[s[k][k]][k[s[k]]]]]]]]][s[k[k]]][s[s[s[s[s[k][k]][k[k]]][k[s[k]]]][s[s[s[
s[k][k]][k[k]]][k[k]]][k[s[k]]]]]][s[s[s[s[s[k][k]][k[k]]][k[k]]][k[s[k]]]][k[s[k]]]][s[s[s[s[s[k][
k]][k[k]]][k[k]]][k[k]]][k[s[k]]]][k[s[k]]][k[k]]]]]]]]]]]][s[s[k[s]]][s[k[s[s[k][k]]]]][s[k[k]]][s[s[k[s]]][s[
k[s[s[k][k]]]]][s[k[k]]][s[s[k][k]][k[k]]]]]]][k[k[k]]]]]]]][k[k[k]]]]][k[s[k]]]]]]]][k[k]]]]]][s[k[k]]][s[k[s[s[
k[s]]][k]]]]][s[s[k][k]][k[s[k]]]]]]]
```

A combinator expression that corresponds to the operation of doing one step of rule 110 evolution.

There has in the past been no overall context for understanding universality in combinators. But now what we have seen suggests that such universality is in a sense just associated with general complex behavior.

Yet we saw [on pages 307–313] that there are symbolic systems with rules even simpler than combinators that still show complex behavior. And so now I suspect that these too are universal.

And in fact wherever one looks, the threshold for universality seems to be much lower than one would ever have imagined. And this is one of the important basic observations that led me to formulate the Principle of Computational Equivalence that I discuss in [Chapter 12 of *A New Kind of Science*].

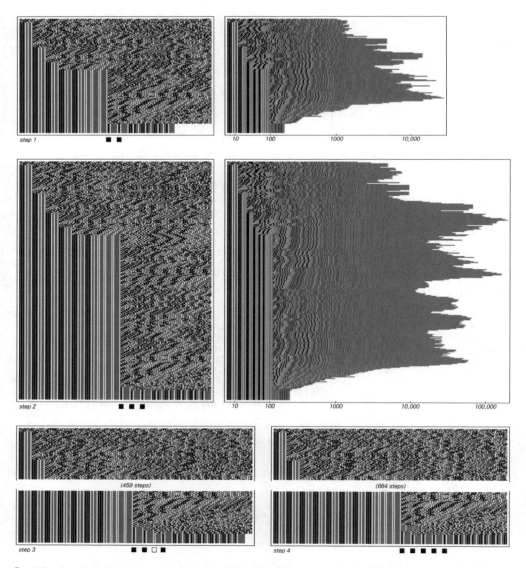

Emulating the rule 110 cellular automaton using combinators. The rule 110 combinator from the previous page is applied once for each step of rule 110 evolution. The initial state is taken to consist of a single black cell.

NOTES

Combinators. After it became widely known in the 1910s that *Nand* could be used to build up any expression in basic logic Moses Schönfinkel introduced combinators in 1920 with the idea of providing an analogous way to build up functions—and to remove any mention of variables—particularly in predicate logic (see page [313]). Given the combinator rules

crules = {s[x_][y_][z_] → x[z][y[z]], k[x_][y_] → x}

the setup was that any function *f* would be written as some combination of *s* and *k*—which Schönfinkel referred to respectively as "fusion" and "constancy"—and then the result of applying the function to an argument *x* would be given by *f[x] //. crules*. (Multiple arguments were handled for example as *f[x][y][z]* in what became known as "currying".) A very simple example of a combinator is *id = s[k][k]*, which corresponds to the identity function, since *id[x] //. crules* yields for any *x*. (In general any combinator of the form *s[k][_]* will also work.) Another example of a combinator is *b = s[k[s]][k]*, for which *b[x][y][z] //. crules* yields *x[y[z]]*.

With the development of lambda calculus in the early 1930s it became clear that given any expression *expr* such as *x[y[x][z]]* with a list of variables *vars* such as *{x, y, z}* one can always find a combinator equivalent to a lambda function such as *Function[x, Function[y, Function[z, x[y[x][z]]]]]*, and it turns out that this can be done simply using

ToC[expr_, vars_] := Fold[rm, expr, Reverse[vars]]
rm[v_, v_] = id
rm[f_[v_], v_] /; FreeQ[f , v] = f
rm[h_, v_] /; FreeQ[h, v] = [h]
rm[f_[g_], v_] := [rm[f , v]][rm[g, v]]

So this shows that any lambda function can in effect be written in terms of combinators, without anything analogous to variables ever explicitly having to be introduced. And based on the result that lambda functions can represent recursive functions, which can in turn represent Turing machines (see note above), it has been known since the mid-1930s that combinators are universal. The rule 110 combinator on page [317] provides however a much more direct proof of this.

The usual approach to working with combinators involves building up arithmetic constructs from them. This typically begins by using so-called Church numerals (based on work by Alonzo Church on lambda calculus), and defining a combinator e_n to correspond to an integer *n* if e_n*[a][b] //. crules* yields *Nest[a, b, n]*. (The *e* on page [308] can thus be considered a Church numeral for 2 since *e[a][b]* is *a[a[b]]*.) This can be achieved by taking e_n to be where

zero = s[k]
inc = s[s[k[s]][k]]

With this setup one then finds

plus = s[k[s]][s[k[s[k[s]]]][s[k[k]]]]
times = s[k[s]][k]
power = s[k[s[s[k][k]]]][k]

(Note that *power[x][y] //. crules* is *y[x]*, and that by analogy *x[x[y]]* corresponds to y^{x^2}, *x[y[x]]* to x^{xy}, *x[y][x]* to x^{y^x}, and so on.)

Another approach involves representing integers directly as combinator expressions. As an example, one can take *n* to be represented just by *Nest[s, k, n]*. And one can then convert any Church numeral *x* to this representation by applying *s[s[s[k][k]][k[s]]][k[k]]*. To go the other way, one uses the result that for all Church numerals *x* and *y*, *Nest[s, k, n][x][y]* is also a Church numeral—as can be seen recursively by noting its equality to *Nest[s, k, n - 1][y][x[y]]*, where as above *x[y]* is *power[y][x]*. And from this it follows that *Nest[s, k, n]* can be converted to the Church numeral for *n* by applying

s[s[s[s[s[k][k]][k[s[s[k[s]][k]][s[k][k]]]]][k[s[s[k[s]][k]][s[s[k[s]][k]][s[k][k]]]]]][s[s[k[s]][s[s[k[s]][s[k
[s[s[s[s[s[s[k][k]][k[s]]][k[k]]][k[s[s[k[s]][k]][s[k][k]]]]][k[s[s[k[s]][k]][s[s[k[s]][k]][s[k][k]]]]]][k[s[
s[s[k][k]][k[s[s[k[s]][s[k[s[s[k][k]]]][s[k][k]]][s[k][k]]][s[s[s[k[s]][k]]][s[s[k[k]][k[k]]]]]][s[k][k]]][
k[k]]]]]][k[s[s[s[k][k]][k[s[k]]][k[s[k][k]]]][k[s[k]]]]]]]][s[k][k]][s[s[k][k]][k[s[s[k][k]][k[s[k][k]]]]][k[s
[s[k[s]][k]][s[s[k[s]][k]][s[k][k]]]]]]]]][k[s[k[k]][s[s[k[s]][k]]]]]]]][k[s[k][k]]]]][k[s[k]]]

Using this one can find from the corresponding results for Church numerals combinator expressions for *plus*, *times* and *power*—with sizes 377, 378 and 382 respectively. It seems certain that vastly simpler combinator expressions will also work, but searches indicate that if *inc* has size less than 4, *plus* must have size at least 8. (Searches based on other representations for integers have also not yielded much. With *n* represented by *Nest[k, s[k][k], n]*, however, *s[s[s[s]][k]][k]* serves as a decrement function, and with *n* represented by *Nest[s[s], s[k], n]*, *s[s[s][k]][k[k[s[s]]]]* serves as a doubling function.

Combinator properties. The size of a combinator expression is conveniently measured by its *LeafCount*. If the evolution of a combinator expression reaches a fixed point, then the expression generated is always the same (Church–Rosser property). But the behavior in the course of the evolution can depend on how the combinator rules are applied; here *expr /. crules* is used at each step, as in the symbolic systems of page [310]. The total number of combinator expressions of successively greater sizes is *{2, 4, 16, 80, 448, 2688, 16 896, 109 824, ...}* (or in general 2^n *Binomial[2 n - 2, n - 1]/ n*; see page [311]). Of these, *{2, 4, 12, 40, 144, 544, 2128, 8544, ...}* are themselves fixed points. Of combinator expressions up to size 6 all evolve to fixed points, in at most *{1, 1, 2, 3, 4, 7}* steps respectively (compare case (a)); the largest fixed points have sizes *{1, 2, 3, 4, 6, 10}* (compare case (b)). At size 7, all but 2 of the 16,896 possible combinator expressions evolve to fixed points, in at most 12 steps (case (c)). The largest fixed point has size 41 (case (d)). *s[s[s]][s[s]][s][s]* (case (e)) and *s[s][s][s[s]][s][s]* lead to expressions that grow like $2^{t/2}$. The maximum number of levels in these expressions (see page [311]) grows roughly linearly, although *Depth[expr]* reaches 14 after 26 and 25 steps, then stays there. At size 8, out of all 109,824 combinator expressions it appears that 49 show exponential growth, and many more show roughly linear growth. *s[s][k][s[s[s]][s]][s]* goes to a fixed point of size 80. *s[s[s]][s[s]][s][s[k]]* (case (i)) increases rapidly to size 7050 but then repeats with period 3. *s[s[s][s]][s]][s][k]* (case (j)) grows to a maximum size of 1263, but then after 98 steps evolves to a fixed point of size 17. For *s[s][k][s[s[s][k]]][k]* (case (k)) the size at step *t - 7* is given by

$h[1] = h[2] = h[3] = 12$

$h[t\_] := If[Mod[t, 4] == 2, 2, 1] (h[Ceiling[t/2] - 1] + t) +$
$\quad \{3, 5, -7, -1\}[[Mod[t, 4] + 1]]$

Examples with similar behavior are *s[s[s][k]][s[s][s[s][k]]*, *s[s[s]][s[s][s[s][k]][k]* and *s[s[s]][s]][s][s[s][k]]*. Among those with roughly exponential growth but seemingly random fluctuations are *s[s[s[s]]][s]][s][k]*, *s[s[s]][s][s[s][s]][k]* and *s[s[s[s]]][s][s]][k][s]*.

Single universal combinators. As already noted by Moses Schönfinkel in 1920, it is possible to set up combinator systems with just a single combinator. In such cases, combinator expressions can be viewed as binary trees without labels, equivalent to balanced strings of parentheses (see page 989) or sequences of 0's and 1's. One example of a single combinator system can be found using *{s → j[j], k → j[j[j]]}*, and has combinator rules (whose order matters):

{j[j][x_][y_][z_] → x[z][y[z]], j[j[j]][x_][y_] → x}

The smallest initial conditions in this case that lead to unbounded growth are of size 14; two are versions of those for *s, k* combinators above, while the third is *j[j][j[j[j]]][j[j]][j[j]][j[j][j[j]]][j[j]][j[j]].*

The forms *j[j]* and *j[j[j]]* appear to be the simplest that can be used for *s* and *k*; and *j[j]* , for example, do not work.

Cellular automaton combinators. With *k* and *s[k]* representing respectively cell values *0* and *1*, a combinator *f* for which *f[a_1][a_0][a_1]* gives the new value of a single cell in an elementary cellular automaton with rule number *m* can be constructed as

Apply[p[p[p[#1][#2]][p[#3][#4]]][p[p[#5][#6]][p[#7][
#8]]] /. {0 → k, 1 → s[k]} &, IntegerDigits[m, 2, 8]] //. crules

where

p = ToC[z[y][x], {x, y, z}] //. crules

The resulting combinator has size 61, but for specific rules somewhat smaller combinators can be found—an example for rule 90 is *s[k[k]][s[s][k[s[s[s[k][k]][k[s[k]]]][k[k]]]]]* with size 16.

To emulate cellular automaton evolution one starts by encoding a list of cell values by the single combinator

p[num[Length[list]]][Fold[p[Nest[s, k, #2]][#1] &, p[k][k], list]] //. crules

where

num[n_] := Nest[inc, s[k], n]
nc = s[s[k[s]][k]]

One can recover the original list by using

Extract[expr, Map[Reverse[IntegerDigits[#, 2]] &,
3 + 59(16^Range[Depth[expr[s[k]][s]][k] //. crules] -
1, 1, -1] - 1)/ 15]]]/. {k → 0, s[k] → 1}

In terms of the combinator *f* a single complete step of cellular automaton evolution can be represented by

w = cr[p[inc[inc[x[s[k]]]]][
inc[x[s[k]]]][cr[p[y[s[k]][k]][p[y[s[k]]][s[k]]][y[k]]],
{y}]][p[x[s[k]][cr[p[p[f[y[k]][k][k][s[k]]][y[k][k][s[k]]][
y[k]][s[k]]][y[s[k]]][y[k][k], {y}]][
p[p[k][k]][p[k][x[k]]][s[k]][p[k][p[k][k]]][k]], {x}]
cr[expr_, vars_] := ToC[expr //. crules, vars]

where there is padding with *0* on either side. With this setup *t* steps of evolution are given simply by *Nest[w, init, t]*. With an initial condition of *n* cells, this then takes roughly $(100 + 35 n) t + 33 t^2$ steps of combinator evolution.

Testing universality in symbolic systems. One can tell that a symbolic system is universal if one can find expressions that act like the *s* and *k* combinators, so that, for example, for some expression *e, e[x][y][z]* evolves to *x[z][y[z]]*.

Über die Bausteine der mathematischen Logik.

Von

M. Schönfinkel in Moskau.

§ 1.

Es entspricht dem Wesen der axiomatischen Methode, wie sie heute vor allem durch die Arbeiten Hilberts zur Anerkennung gelangt ist, daß man nicht allein hinsichtlich der Zahl und des Gehalts der Axiome nach möglichster Beschränkung strebt, sondern auch die Anzahl der als unentbehrlich zugrunde zu legenden Begriffe so klein wie möglich zu machen sucht, indem man nach Begriffen fahndet, die vorzugweise geeignet sind, um aus ihnen alle anderen Begriffe des fraglichen Wissenszweiges auf ...

An Analysis of Logical Substitution.

By H. B. CURRY.

Contents.*

1. Preliminary Discussion of the Nature of Mathematical Logic.
2. Logical Substitution; its Relation to a Combinatory Problem.
3. Solution of the Combinatory Problem.

Mathematical Logic has been defined as an application of the methods of mathematics to the domain of Logic. In accordance with the analysis and criticism of the logic, to the construction of ...

To H.B. Curry: Essays on
Combinatory Logic, Lambda
Calculus and Formalism

Edited by
J.P. SELDIN
Mathematics Department

THE ρ-FUNCTION IN λ-K-CONVERSION

A. M. TURING

In the theory of conversion it is important to have a formally defined function which assigns to any positive integer n the least integer not less than n which has a given property. The definition of such a formula is somewhat involved ...

A BASIS RESULT IN COMBINATORY LOGIC

REMI LEGRAND

The aim of this article is to show that a basis for combinatory logic [2] must contain at least one combinator with rank strictly greater than two. We use notation of [1].

DEFINITION 1. Let Q be a primitive combinator given by its reduction rule $Qx_1...x_n \to C$, where C is a pure combination of the variables $x_1,...,x_n$ is called the rank of the combinator.

DEFINITION 2. A set $\{Q_1,...,Q_k\}$ of combinators is a basis for combinatory logic if for every finite set $\{x_1,...,x_n\}$ of variables and every pure combination of these variables, there exists a pure combination Q of $Q_1,...,Q_k$ such that $Qx_1...x_n \to C$.

Eidgenössische
Technische
Hochschule
Zürich

Berichte
des
Instituts für
Informatik

Efstathios Zachos

Kombinatorische
Logik und
S-Terme

Problem Corner

A Case Study in Automated Theorem Proving:
Finding Sages in Combinatory Logic*

WILLIAM McCUNE and LARRY WOS

1. Introduction

Ross Overbeek recently brought to our attention a problem in combinatory logic.

ROBERT FEYS

PEANO ET BURALI-FORTI PRECURSEURS DE LA LOGIQUE
COMBINATOIRE

The Systematic Construction of a
One-Combinator Basis for Lambda-Terms

Jeroen Fokker

Department of Computer Science, Utrecht University, The Netherlands

Keywords: SKI-combinators; Systematic derivation; Simplicity as objective.

AN ABSTRACTION ALGORITHM FOR COMBINATORY LOGIC

S. KAMAL ABDALI

STRONG REDUCTION AND NORMAL FORM IN
COMBINATORY LOGIC

BRUCE LERCHER

The notion of strong reduction is introduced in Curry and Feys' book Combinatory logic [1] as an analogue, in the theory of combinators, to reduction (more exactly, βη-reduction) in the theory of λ-conversion.

COMBINATORY REDUCTIONS AND LAMBDA REDUCTIONS COMPARED

by ROGER HINDLEY in Swansea (Great Britain)

Introduction

This paper is an informal comparison of some of the basic properties of combinatory reductions and lambda reductions.

EIN ALGEBRAISCHER BEWEIS
FÜR DAS CHURCH-ROSSER-THEOREM*

Von Gerd Mitschke in Darmstadt

§0 Einleitung

Der wichtigste Satz der Theorien der λ-Konversion und nach der kombinatorischen Logik ist zweifellos das sog. "Church-Rosser-Theorem" (Church-Rosser [2]).

The Logical Way of
Doing Things

edited by Karel Lambert

COMPACT NUMERAL REPRESENTATION WITH COMBINATORS

R. V. KRISHNAMURTHY and R. F. VICKERS

A Bibliography of Combinators

Foundational Documents

M. Schönfinkel (1924), "Über die Bausteine der mathematischen Logik" (in German) ["On the Building Blocks of Mathematical Logic"], *Mathematische Annalen* 92, 305–316. doi: 10.1007/BF01448013. (Translated by S. Bauer-Mengelberg (1967), as "On the Building Blocks of Mathematical Logic", in *From Frege to Gödel: A Source Book in Mathematical Logic, 1879–1931*, J. van Heijenoort, Harvard University Press, 357–366.)

H. B. Curry (1929), "An Analysis of Logical Substitution", *American Journal of Mathematics* 51, 363–384. doi: 10.2307/2370728.

H. B. Curry (1930), "Grundlagen der kombinatorischen Logik" (in German) ["Foundations of Combinatory Logic"], *American Journal of Mathematics* 52, 509–536. doi: 10.2307/2370619. (Translated by F. Kamareddine and J. Seldin (2016), as *Foundations of Combinatory Logic,* Logic PhDs vol. 1, College Publications.)

Books

H. B. Curry, R. Feys and W. Craig (1958), *Combinatory Logic*, North-Holland.

H. Barendregt (1971), *Some Extensional Term Models for Combinatory Logics and Lambda Calculi*, Amsterdam. Republished 2006.

H. B. Curry, J. R. Hindley and J. P. Seldin (1972), *Combinatory Logic, Volume II*, North Holland.

S. Stenlund (1972), *Combinators, λ-Terms and Proof Theory*, D. Reidel Publishing Co.

F. B. Fitch (1974), *Elements of Combinatory Logic*, Yale University Press.

J. R. Hindley and J. P. Seldin (eds.) (1980), *To H. B. Curry: Essays on Combinatory Logic, Lambda Calculus and Formalism*, Academic Press, Inc.

J. W. Klop (1980), *Combinatory Reduction Systems*, Mathematisch Centrum.

H. P. Barendregt (1981), *The Lambda Calculus: Its Syntax and Semantics*, North-Holland. Second edition 1984, republished 2006.

R. Smullyan (1985), *To Mock a Mockingbird,* Knopf.

J. R. Hindley and J. P. Seldin (1986), *Introduction to Combinators and Lambda-Calculus,* Cambridge University Press.

G. E. Révész (1988), *Lambda-Calculus, Combinators, and Functional Programming*, Cambridge University Press.

C. L. Hankin (1994), *Lambda Calculi: A Guide for Computer Scientists*, Clarendon Press.

E. Engeler (1995), *The Combinatory Programme*, Springer.

V. E. Wolfengangen (2003), *Combinatory Logic in Programming: Computations with Objects through Examples and Exercises*, Center JurInfoR Ltd. archive.org/details/CLP-2003_780.

J. R. Hindley and J. P. Seldin (2008), *Lambda-Calculus and Combinators: An Introduction*, Cambridge University Press. doi: 10.1017/CBO9780511809835.

K. Bimbó (2011), *Combinatory Logic: Pure, Applied and Typed*, Chapman and Hall/CRC. doi: 10.1201/b11046.

H. Barendregt (2020), *Some Extensional Term Models for Combinatory Logics and Lambda Calculi 1971 (Series on Term Rewriting and Logic): A 2020 Republication: Motivation, the Making of & Hindsight*, independently published.

Surveys & Summaries

P. C. Rosenbloom (1950), "III.4: Combinatory Logics", in *The Elements of Mathematical Logic*, Dover, 109–133.

H. B. Curry (1955), "La théorie des combinateurs" (in French), *Rendiconti di Matematica e delle sue Applicazioni* 10, 347–359.

H. B. Curry (1957), "Combinatory Logic", in *Summaries of Talks Presented at the Summer Institute of Symbolic Logic at Cornell University, July 1957*, notes by J. B. Rosser, Communications Research Division, Institute for Defense Analyses, 90–99.

H. B. Curry (1967), "Logic, Combinatory", *The Encyclopedia of Philosophy*, Volume 4, R. Edwards (ed.), Macmillan and Free Press, 504–509.

W. V. Quine (1967), "Introduction to 'Bausteine der mathematischen Logik'", in *From Frege to Gödel: A Source Book in Mathematical Logic, 1879–1931*, J. van Heijenoort, Harvard University Press, 355–357.

M. W. Bunder (1972), "An Introduction to, and Trends in Combinatorial Logic", in *Proceedings of the First Australian Conference on Combinatorial Mathematics*, J. Wallis and W. D. Wallis (eds.), Australian Conference on Combinatorial Mathematics, TUNRA, 183–191.

G. W. Petznick (1974), "Introduction to Combinatory Logic" in *Theory of Computation*, W. S. Brainerd and L. H. Landweber, Wiley & Sons, 274–323.

H. B. Curry (1980), "Some Philosophical Aspects of Combinatory Logic", *Studies in Logic and the Foundations of Mathematics* 101, 85–101. doi: 10.1016/S0049-237X(08)71254-0.

A. Rezus (1982), *A Bibliography of Lambda-Calculi, Combinatory Logic and Related Topics*, Mathematisch Centrum. researchgate.net/publication/239844967_A_bibliography_of _lambda-calculi_combinatory_logics _and_related_topics.

J. R. Hindley (1986), "Combinators and Lambda-Calculus, a Short Outline", in *Combinators and Functional Programming Languages*, G. Cousineau, P.-L. Curien and B. Robinet (eds.), LITP Spring School on Theoretical Computer Science 1985, *Lecture Notes in Computer Science* vol. 242, Springer, 104–122. doi: 10.1007/3-540-17184-3_42.

R. D. Knott (1988), "Introduction to Combinators and Lambda-Calculus", *University Computing* 10, 166–167.

P. Freyd (1989), "Combinators", *Categories in Computer Science and Logic,* J. W. Gary and A. Scedrov (eds.), AMS, 63–66. doi: 10.1090/conm/092/1003195.

J. W. Klop, V. van Oostrom and F. van Raamsdonk (1993), "Combinatory Reduction Systems: Introduction and Survey", *Theoretical Computer Science* 121, 279–308. doi: 10.1016/0304-3975(93)90091-7.

F. Cardone and J. R. Hindley (2006), "History of Lambda-Calculus and Combinatory Logic", Swansea University Mathematics Department Research Report No. MRRS-05-06. cs.vu.nl/~femke/courses/ep/slides/history.pdf.

K. Bimbó (2009), "Combinatory Logic", *Stanford Encyclopedia of Philosophy.* plato.stanford.edu/entries/logic-combinatory.

F. Cardone and J. R. Hindley (2009), "Lambda-Calculus and Combinators in the 20th Century", in *Logic from Russell to Church*, D. M. Gabbay and J. Woods (eds.), Elsevier, 723–817. doi: 10.1016/S1874-5857(09)70018-4.

Combinators as Symbolic Expressions

Specific Combinators & Behavior

A. M. Turing (1937), "The þ-Function in λ-K-Conversion", *The Journal of Symbolic Logic* 2, 164. doi: 10.2307/2268281.

F. B. Fitch (1969), "Combinatory Logic and Negative Numbers", in *The Logical Way of Doing Things, in Honor of Henry S. Leonard,* K. Lambert (ed.), Yale University Press, 265–277.

H. Barendregt and J. B. Rosser (1971), letter correspondence. wolframcloud.com/obj/combinators/Bibliography/Barendregt-Rosser-LetterCorrespondence.

A. Mitschke and G. Mitschke (1975), "A Solution of Böhm's S-Problem", Technische Hochschule Darmstadt, unpublished manuscript. wolframcloud.com/obj/combinators/Bibliography/Mitschke-Mitschke-ASolutionofBohmsSProblem.

A. S. Kuzičev (1976), "Combinatorially Complete Systems with Operators Ξ, F, Q, Π, $\exists$, P, $\neg$, &, $\vee$, $\equiv$, .", *Moscow University Mathematics Bulletin* 31, 117–122. istina.msu.ru/publications/article/18961167.

S. Zachos (1978), "Kombinatorishe Logik und-S-Terme" (in German), dissertation, ETH Zürich. doi: 10.3929/ethz-a-000136021.

J. Shultis (1985), "A Note on the S Combinator", CU-CS-304-85. core.ac.uk/download/pdf/54846325.pdf.

W. McCune and L. Wos (1987), "A Case Study in Automated Theorem Proving: Finding Sages in Combinatory Logic". *Journal of Automated Reasoning* 3, 91–108. doi: 10.1007/BF00381147.

R. Legrand (1988), "A Basis Result in Combinatory Logic", *The Journal of Symbolic Logic* 53, 1224–1226. doi: 10.2307/2274616.

L. Wos and W. McCune (1988), "Searching for Fixed Point Combinators by Using Automated Theorem Proving: A Preliminary Report", Report ANL-88-10, Argonne National Laboratory, Argonne, IL. doi: 10.2172/6852789.

J. D. Fokker (1989), *The Systematic Construction of a One-Combinator Basis*, University of Utrecht. cs.uu.nl/research/techreps/repo/CS-1989/1989-14.pdf.

R. Statman (1989), "The Word Problem for Smullyan's Lark Combinator Is Decidable", *Journal of Symbolic Computation* 7, 103–112. doi: 10.1016/S0747-7171(89)80044-6.

R. Statman (1989), "On Sets of Solutions to Combinator Equations", *Theoretical Computer Science* 66, 99–104. doi: 10.1016/0304-3975(89)90148-5.

N. Dershowitz and R. Treinen (1991, accessed April 9, 2021), "The RTA List of Open Problems". win.tue.nl/rtaloop/problems/summary.html.

W. McCune and L. Wos (1991), "The Absence and the Presence of Fixed Point Combinators", *Theoretical Computer Science* 87, 221–228. doi: 10.1016/0304-3975(91)90034-Y.

R. Statman (1991), "There Is No Hyperrecurrent S, K Combinator", Carnegie Mellon University. Journal contribution. doi: 10.1184/R1/6480071.v1.

J. Fokker (1992), "The Systematic Construction of a One-Combinator Basis for Lambda-Terms", *Formal Aspects of Computing* 4, 776–780. doi: 10.1007/BF03180572.

M. Sprenger and M. Wymann-Böni (1993), "How to Decide the Lark", *Theoretical Computer Science* 110, 419–432. doi: 10.1016/0304-3975(93)90015-L.

R. Statman (1993), "Some Examples of Non-existent Combinators", *Theoretical Computer Science* 121, 441–448. doi: 10.1016/0304-3975(93)90096-C.

L. Wos (1993), "The Kernel Strategy and Its Use for the Study of Combinatory Logic", *Journal of Automated Reasoning* 10, 287–343. doi: 10.1007/BF00881795.

P. Trigg, J. R. Hindley and M. W. Bunder (1994), "Combinatory Abstraction Using **B**, **B′** and Friends", *Theoretical Computer Science* 135, 405–422. doi: 10.1016/0304-3975(94)90114-7.

B. Intrigila (1997), "Non-existent Statman's Double Fixed Point Combinator Does Not Exist, Indeed", *Information and Computation* 137, 35–40. doi: 10.1006/inco.1997.2633.

R. Statman (2000), "On the Word Problem for Combinators", in *Rewriting Techniques and Applications*, L. Bachmair (ed.), International Conference on Rewriting Techniques and Applications 2000, *Lecture Notes in Computer Science* vol. 1833, Springer, 203–213. doi: 10.1007/10721975_14.

J. Waldmann (2000), "The Combinator **S**", *Information and Computation* 159, 2–21. doi: 10.1006/inco.2000.2874.

D. Probst and T. Studer (2001), "How to Normalize the Jay", *Theoretical Computer Science* 254, 677–681. doi: 10.1016/S0304-3975(00)00379-0.

T. Dörges (2002), "Unendliche Reduktionen in der kombinatorischen Logik" (in German). Diplomarbeit, Universitat Leipzig. nbn-resolving.org/urn:nbn:de:bsz:15-qucosa2-165217.

B. J. Maclennan (2002), "Molecular Combinator Reference Manual", Department of Computer Science, University of Tennessee, Knoxville, UPIM Report 2, Technical Report UT-CS-02-489. researchgate.net/publication/2854487_Molecular_Combinator_Reference_Manual.

S. Wolfram (2002), "Symbolic Systems", etc. in *A New Kind of Science*, Wolfram Media, wolframscience.com/nks [included in this book].

R. Statman (2005), "Two Variables Are Not Enough", in *Theoretical Computer Science*, M. Coppo, E. Lodi and G. M. Pinna (eds.), Italian Conference on Theoretical Computer Science 2005, *Lecture Notes in Computer Science* vol. 3701, Springer, 406–409. doi: 10.1007/11560586_32.

J. W. Klop (2007), "New Fixed Point Combinators from Old", in "Reflections on Type Theory, λ-Calculus, and the Mind. Essays Dedicated to Henk Barendregt on the Occasion of his 60th Birthday", E. Barendsen, et al. (eds.), University Nijmegen, 197–210. cs.ru.nl/barendregt60/essays/klop/art16_klop.pdf.

J. Endrullis, R. de Vrijer and J. Waldmann (2010), "Local Termination: Theory and Practice", *Logical Methods in Computer Science* 6, 1–37. doi: 10.2168/LMCS-6(3:20)2010.

P. Cheilaris, J. Ramirez and S. Zachos (2011), "Checking in Linear Time if an *S*-Term Normalizes", in *Proceedings of the 8th Panhellenic Logic Symposium,* Panhellenic Logic Symposium 8. 147.102.36.12/~philaris/publications/PLS8_submission_23.pdf.

H. P. Barendregt, et al. (2017), "Dance of the Starlings", in *Raymond Smullyan on Self Reference*, M. Fitting and B. Rayman (eds.), "Outstanding Contributions to Logic" vol. 14, Springer, 67–111. doi: 10.1007/978-3-319-68732-2_5.

M. Ikebuchi and K. Nakano (2018), "On Repetitive Right Application of B-Terms". arXiv:1703.10938.

E. Paul (2020), "Analyzing and Computing with Combinators", Wolfram Summer Camp 2020. community.wolfram.com/groups/-/m/t/2034214.

Conversions & Notations

H. B. Curry (1933), "Apparent Variables from the Standpoint of Combinatory Logic", *Annals of Mathematics Second Series* 34, 381–404. jstor.org/stable/1968167.

W. V. Quine (1936), "A Reinterpretation of Schönfinkel's Logical Operators", *Bulletin of the American Mathematical Society* 42, 87–89. Zbl: 0014.00301.

R. Feys (1946), "La technique de la logique combinatoire" (in French), *Revue philosophique de Louvain* 44, 74–103. jstor.org/stable/26332881.

R. Feys (1953), "Peano et Burali-Forti précurseurs de la logique combinatoire" (in French), *Actes du XIème Congrès International de Philosophie* 5, 70–72. doi: 10.5840/wcp1119535131.

F. B. Fitch (1957), "Combinatory Logic and Whitehead's Theory of Prehensions", *Philosophy of Science* 24, 331–335. jstor.org/stable/185007.

H. B. Curry (1964), "The Elimination of Variables by Regular Combinators 1", in *The Critical Approach to Science and Philosophy: Essays in Honor of Karl Popper*, M. Bunge (ed.), The Free Press, 127–143. Republished 1999 and 2017. doi: 10.4324/9781351313087-9.

D. Meredith (1974), "Combinatory and Propositional Logic", *Notre Dame Journal of Formal Logic* 15, 156–160. doi: 10.1305/ndjfl/1093891208.

M. A. Amer (1975), "Parentheses in Combinatory Logic", *Studies in Logic and the Foundations of Mathematics* 80, 429–432. doi: 10.1016/S0049-237X(08)71961-X.

D. Meredith (1975), "Combinator Operations", *Studia Logica* 24, 367–385. doi: 10.1007/BF02121666.

D. Scott (1975), "Some Philosophical Issues Concerning Theories of Combinators", in *λ-Calculus and Computer Science Theory*, C. Böhm (ed.), International Symposium on Lambda-Calculus and Computer Science Theory 1975, *Lecture Notes in Computer Science* vol. 37, Springer, 346–366. doi: 10.1007/BFb0029537.

S. K. Abdali (1976), "An Abstraction Algorithm for Combinatory Logic", *Journal of Symbolic Logic* 41, 222–224. jstor.org/stable/2272961.

R. Canal (1978), "Complexité de la réduction en logique combinatoire" (in French), dissertation, Université Paul Sabatier.

D. A. Turner (1979), "Another Algorithm for Bracket Abstraction", *The Journal of Symbolic Logic* 44, 267–270. doi: 10.2307/2273733.

B. Robinet (1982), "Sur des séquences itératives de combinateurs" (in French) ["On Iterative Sequences of Combinators"], *Comptes Rendus de l'Academie des Sciences de Paris* 295, 29–30.

E. V. Krishnamurthy and B. P. Vickers (1987), "Compact Numeral Representation with Combinators", *The Journal of Symbolic Logic* 52, 519–525. doi: 10.2307/2274398.

A. Piperno (1989), "Abstraction Problems in Combinatory Logic: A Compositive Approach", *Theoretical Computer Science* 66, 27–43. doi: 10.1016/0304-3975(89)90143-6.

S. Broda and L. Damas (1995), "A New Translation Algorithm from Lambda Calculus into Combinatory Logic", in *Progress in Artificial Intelligence*, C. Pinto-Ferreira and N. J. Mamede (eds.), Portuguese Conference on Artificial Intelligence 1995, *Lecture Notes in Computer Science* vol. 990, Springer, 359-370. doi: 10.1007/3-540-60428-6_30.

D. C. Keenan (1996), "To Dissect a Mockingbird: A Graphical Notation for the Lambda Calculus with Animated Reduction". dkeenan.com/Lambda.

S. Broda and L. Dama (1997), "Compact Bracket Abstraction in Combinatory Logic", *The Journal of Symbolic Logic* 62, 729–740. doi: 10.2307/2275570.

C. Okasaki (2003), "Flattening Combinators: Surviving without Parentheses", *Journal of Functional Programming* 13, 815–822. doi: 10.1017/S0956796802004483.

J. Tromp (2004), "Binary Lambda Calculus", GitHub. tromp.github.io/cl/Binary_lambda_calculus.html.

J. Tromp (2007), "Binary Lambda Calculus and Combinatory Logic", *Randomness and Complexity, From Leibniz to Chaitin*, 237–260. doi:10.1142/9789812770837_0014.

B. Jay and T. Given-Wilson (2011), "A Combinatory Account of Internal Structure", *The Journal of Symbolic Logic* 76, 807–826. doi: 10.2178/jsl/1309952521.

Combinator Reduction

H. B. Curry (1932), "Some Additions to the Theory of Combinators", *American Journal of Mathematics* 54, 551–558. doi: 10.2307/2370900.

K. L. Loewen (1962), "A Study of Strong Reduction in Combinatory Logic", dissertation, the Pennsylvania State University.

B. Lercher (1963), "Strong Reduction and Recursion in Combinatory Logic", dissertation, the Pennsylvania State University.

B. Lercher (1967), "Strong Reduction and Normal Form in Combinatory Logic", *The Journal of Symbolic Logic* 32, 213–223. doi: 10.2307/2271659.

K. L. Loewen (1968), "Modified Strong Reduction in Combinatory Logic", *Notre Dame Journal of Formal Logic* 9, 265–270. doi: 10.1305/ndjfl/1093893461.

K. L. Loewen (1968), "The Church Rosser Theorem for Strong Reduction in Combinatory Logic", *Notre Dame Journal of Formal Logic* 9, 299–302. doi: 10.1305/ndjfl/1093893514.

G. Mitschke (1972), "Ein algebraischer Beweis für das Church-Rosser-Theorem" (in German), *Archiv für mathematische Logik und Grundlagenforschung* 15, 146–157. doi: 10.1007/BF02008531.

C. Batini and A. Pettorossi (1975), "Some Properties of Subbases in Weak Combinatory Logic", Istituto di Automatica, University of Rome report 75-04. wolframcloud.com/obj/combinators/Bibliography/Batini-Pettorossi-SomePropertiesofSubbasesinWeakCombinatoryLogic.

A. Pettorossi (1975), "Sulla terminazione in classi subrecursive di algoritmi" (in Italian) ["On Termination in Subrecursive Classes of Algorithms"], in *Proceedings AICA Congress,* International Association for Analog Computation Congress 1975, 62–67. wolframcloud.com/obj/combinators/Bibliography/Pettorossi-OnTerminationinSubrecursiveClassesofAlgorithms.

A. Pettorossi (1975), "Un teorema per la sottobase {B}" (in Italian) ["A Theorem for the Subbase {B}]", Convegno su: "Codici, Complessità di Calcolo e Linguaggi Formali" 1975, Laboratorio di Cibernetica del CNR, 62-67. wolframcloud.com/obj/combinators/Bibliography/Pettorossi-ATheoremfortheSubbaseB.

J. R. Hindley (1977), "Combinatory Reductions and Lambda Reductions Compared", *Mathematical Logic Quarterly* 23, 169–180. doi: 10.1002/malq.19770230708.

R. J. Lipton and L. Snyder (1977), "On the Halting of Tree Replacement Systems", Department of Computer Science, Yale University, Research Report #99. wolframcloud.com/obj/combinators/Bibliography/Lipton-Snyder-OntheHaltingofTreeReplacementSystems.

G. Mitschke (1977), "Discriminators in Combinatory Logic and λ-Calculus" preprint 336, Darmstadt University. wolframcloud.com/obj/combinators/Bibliography/Mitschke-DiscriminatorsinCombinatoryLogicandLambdacalculus.

A. Pettorossi (1977), "Combinators as Tree-Transducers", *Les arbres en algèbre et en programmation*, G. Jacobs (ed.), Université de Lille.

R. Canal (1978), "Complexité de la réduction en logique combinatorie" (in French), *RAIRO Informatique Théorique* 12, 185–199. doi: 10.1051/ita/1978120403391.

G. Jacopini and M. Venturini-Zilli (1978), "Equating for Recurrent Terms of Lambda Calculus and Combinatory Logic", *Pubblicazioni dell' Istituto per le Applicazioni del Calcolo*, CRN, Ser. III, 85. wolframcloud.com/obj/combinators/Bibliography/Jacopini-Venturini-Zilli-EquatingforRecurrentTermsofLambdaCalculusandCombinatoryLogic.

M. V. Zilli (1978), "Head Recurrent Terms in Combinatory Logic: A Generalization of the Notion of Head Normal Form", in *Automata, Languages and Programming*, G. Ausiello and C. Böhm (eds.), International Colloquium on Automata, Languages, and Programming 1978, *Lecture Notes in Computer Science* vol. 62, Springer, 477–493. doi: 10.1007/3-540-08860-1_36.

R. Canal and J. Vignolle (1979), "Calculs finis et infinis dans les termes combinatoires" (in French), in *Lambda Calcul et Sémantique Formelle des langages de programmation*, B. Robinet (ed.), Litp-Ensta 109–130. wolframcloud.com/obj/combinators/Bibliography/Canal-Vignolle-Calculsfinisetinfinisdanslestermescombinatoires.

G. Mitschke (1979), "The Standardization Theorem for λ-Calculus", *Zeitschrift für mathematische Logik und Grundlagen der Mathematik* 25, 29–31. doi: 10.1002/malq.19790250104.

C. Böhm (1980), "An Abstract Approach to (Hereditary) Finite Sequences of Combinators", in *To H. B. Curry: Essays on Combinatory Logic, Lambda Calculus and Formalism*, Academic Press, 231–242.

J. W. Klop (1980), "Reduction Cycles in Combinatory Logic", in *To H. B. Curry: Essays on Combinatory Logic, Lambda Calculus and Formalism*, J. R. Hindley and J. P. Seldin (eds.), Academic Press, Inc., 193–214.

G. Mitschke (1980), "Infinite Terms and Infinite Reductions", in *To H. B. Curry: Essays on Combinatory Logic, Lambda Calculus and Formalism*, J. R. Hindley and J. P. Seldin (eds.), Academic Press Inc., 243–257.

A. Pettorossi (1981), "A Property Which Guarantees Termination in Weak Combinatory Logic and Subtree Replacement Systems", *Notre Dame Journal of Formal Logic* 22, 289–300. doi: 10.1305/ndjfl/1093883514.

L. V. Shabunin (1983), "On the Interpretation of Combinators with Weak Reduction", *The Journal of Symbolic Logic* 48, 558–563. doi: 10.2307/2273446.

L. Wos and W. McCune (1988), "Challenge Problems Focusing on Equality and Combinatory Logic: Evaluating Automated Theorem-Proving Programs", in *9th International Conference on Automated Deduction*, E. Lusk and R. Overbeek (eds.), Conference on Automated Deduction 1988, *Lecture Notes in Computer Science* vol. 310, Springer, 714–729. doi: 10.1007/BFb0012870.

W. M. Farmer, J. D. Ramsdell and R. J. Watro (1990), "A Correctness Proof for Combinator Reduction with Cycles", *ACM Transactions on Programming Languages and Systems* 12, 123–134. doi: 10.1145/77606.77612.

A. Berarducci (1994), "Infinite λ-Calculus and Non-sensible Models", Logic and Algebra, Routledge, 339–377. people.dm.unipi.it/berardu/Art/1996Nonsensible/non-sensible.pdf.

R. Statman (1997), "Effective Reduction and Conversion Strategies for Combinators", in *Rewriting Techniques and Applications*, H. Coman (ed.), International Conference on Rewriting Techniques and Applications 1997, *Lecture Notes in Computer Science* vol. 1232, Springer, 299–307. doi: 10.1007/3-540-62950-5_79.

B. Intrigila and E. Biasone (2000), "On the Number of Fixed Points of a Combinator in Lambda Calculus", *Mathematical Structures in Computer Science* 10, 595–615. doi: 10.1017/S0960129500003091.

P. Minari (2009), "A Solution to Curry and Hindley's Problem on Combinatory Strong Reduction", *Archive for Mathematical Logic* 48, 159–184. doi: 10.1007/s00153-008-0109-z.

A. Charguéraud (2010), "The Optimal Fixed Point Combinator", in *Interactive Theorem Proving*, M. Kaufmann and L. C. Paulson (eds.), International Conference on Interactive Theorem Proving 2010, *Lecture Notes in Computer Science* vol. 6172, Springer, 195–210. doi: 10.1007/978-3-642-14052-5_15.

J. Endrullis, D. Hendriks and J. W. Klop (2010), "Modular Construction of Fixed Point Combinators and Clocked Böhm Trees", in *2010 25th Annual IEEE Symposium on Logic in Computer Science,* IEEE Symposium on Logic in Computer Science 2010, IEEE, 111–119. doi: 10.1109/LICS.2010.8.

P. Dybjer and D. Kuperberg (2012), "Formal Neighbourhoods, Combinatory Böhm Trees, and Untyped Normalization by Evaluation", *Annals of Pure and Applied Logic* 163, 122–131. doi: 10.1016/j.apal.2011.06.021.

B. F. Redmond (2016), "Bounded Combinatory Logic and Lower Complexity", *Information and Computation* 248, 215–226. doi: 10.1016/j.ic.2015.12.013.

J. Endrullis, et al. (2017), "Clocked Lambda Calculus", *Mathematical Structures in Computer Science* 27, 782–806. doi: 10.1017/S0960129515000389.

A. Polonsky (2018), "Fixed Point Combinators as Fixed Points of Higher-Order Fixed Point Generators", *Logical Methods in Computer Science* 16. arXiv:1810.02239.

A. Bhayat and G. Reger (2020), "A Knuth-Bendix-Like Ordering for Orienting Combinator Equations", in *Automated Reasoning,* N. Peltier and V. Sofronie-Stokkermans (eds.), International Joint Conference on Automated Reasoning 2020, *Lecture Notes in Computer Science* vol. 12166, Springer, 259–277. doi: 10.1007/978-3-030-51074-9_15.

J. W. Klop (2020), forthcoming, "An Inkling of Infinitary Rewriting".

Random Combinators

B. J. Maclennan (1997), "Preliminary Investigation of Random SKI-Combinator Trees", Department of Computer Science, University of Tennessee, Knoxville, Technical Report CS-97-370. researchgate.net/publication/2736515_Preliminary_Investigation _of_Random_SKI-Combinator_Trees.

R. David, et al. (2013), "Asymptotically Almost All Lambda-Terms Are Strongly Normalizing", *Logical Methods in Computer Science* 9, 1–30. doi: 10.2168/LMCS-9(1:2)2013.

M. Bendkowski, K. Grygiel and M. Zaionc (2015), "Asymptotic Properties of Combinatory Logic", in *Theory and Applications of Models of Computation,* R. Jain, S. Jain and F. Stephan (eds.), International Conference on Theory and Applications of Models of Computation 2015, *Lecture Notes in Computer Science* vol. 9076, Springer, 62–72. doi: 10.1007/978-3-319-17142-5_7.

M. Bendkowski, K. Grygiel and M. Zaionc (2017), "On the Likelihood of Normalization in Combinatory Logic", *Journal of Logic and Computation* 27, 2251–2269. doi: 10.1093/logcom/exx005.

E. J. Parfitt (2017), "Patterns in Combinator Evolution", *Complex Systems* 26, 119–134. doi: 10.25088/ComplexSystems.26.2.119.

Combinators as Mathematical Constructs

Combinatory Logic

H. B. Curry (1931), "The Universal Quantifier in Combinatory Logic", *Annals of Mathematics Second Series* 32, 154–180. jstor.org/stable/1968422.

H. B. Curry (1934), "Some Properties of Equality and Implication in Combinatory Logic", *Annals of Mathematics Second Series* 35, 849–860. doi: 10.2307/1968498.

H. B. Curry (1934), "Functionality in Combinatory Logic", *Proceedings of the National Academy of Sciences of the United States of America* 20, 584–590. doi: 10.1073/pnas.20.11.584.

H. B. Curry (1935), "First Properties of Functionality in Combinatory Logic", *Tohoku Mathematical Journal* 41, 371–401. jstage.jst.go.jp/article/tmj1911/41/0/41_0_371/_article/-char/en.

J. B. Rosser (1935), "A Mathematical Logic without Variables", *Annals of Mathematics* 36, 127–150. doi: 10.2307/1968669.

H. B. Curry (1941), "A Revision of the Fundamental Rules of Combinatory Logic", *The Journal of Symbolic Logic* 6, 41–53. doi: 10.2307/2266655.

H. B. Curry (1941), "Consistency and Completeness of the Theory of Combinators", *The Journal of Symbolic Logic* 6, 54–61. doi: 10.2307/2266656.

H. B. Curry (1942), "The Combinatory Foundations of Mathematical Logic", *Journal of Symbolic Logic* 7, 49–64. doi: 10.2307/2266302.

J. B. Rosser (1942), "New Sets of Postulates for Combinatory Logics", *The Journal of Symbolic Logic* 7, 18–27. doi: 10.2307/2267551.

H. B. Curry (1948/1949), "A Simplification of the Theory of Combinators", *Synthese* 7, 391–399. jstor.org/stable/20114069.

F. B. Fitch (1960), "A System of Combinatory Logic", Office of Naval Research, Group Psychology Branch, Contract No. SAR/Nonr-609(16), Technical Report.

H. B. Curry (1968), "Recent Advances in Combinatory Logic", *Bulletin de la Société Mathématique Belgique* 20, 288–298.

H. B. Curry (1969), "Modified Basic Functionality in Combinatory Logic", Dialectica 23, 83–92. doi: 10.1111/j.1746-8361.1969.tb01183.x.

A. S. Kuzičev (1971), "Certain Properties of Schönfinkel–Curry Combinators", *Kombinatorny i Anal* 1, 105–119.

H. P. Barendregt (1973), "Combinatory Logic and the Axiom of Choice", *Indagationes Mathematicae* 35, 203–221. doi: 10.1016/1385-7258(73)90005-X.

L. V. Šabunin (1973), "Simple Combinatory Calculi", *Vestnik Moskovskogo Universiteta. Serija I. Matematika, Mehanika* 28, 30–35.

H. P. Barendregt (1974), "Combinatory Logic and the ω-Rule", *Fundamental Mathematicae* 82, 199–215. doi: 10.4064/fm-82-3-199-215.

C. Batini and A. Pettorossi (1975), "On Subrecursiveness in Weak Combinatory Logic" in *λ-Calculus and Computer Science Theory*, C. Böhm (ed.), International Symposium on Lambda-Calculus and Computer Science Theory 1975, *Lecture Notes in Computer Science* vol. 37, Springer, 297–311. doi: 10.1007/BFb0029533.

L. V. Šabunin (1975), "Combinatory Calculi. I." (in Russian), *Vestnik Moskovskogo Universiteta. Serija I. Matematika, Mehanika* 30 (1), 12–17.

L. V. Šabunin (1975), "Combinatory Calculi. II." (in Russian), *Vestnik Moskovskogo Universiteta. Serija I. Matematika, Mehanika* 30 (2), 10–14.

G. Longo (1976), "On the Problem of Deciding Equality in Partial Combinatory Algebras and in a Formal System", *Studia Logica: An International Journal for Symbolic Logic* 35, 363–375. jstor.org/stable/20014826.

J. P. Seldin (1978), "Recent Advances in Curry's Program", *Rendiconti del Seminario Matematico* 35, 77–88.

M. W. Bunder and R. K. Meyer (1985), "A Result for Combinators, BCK Logics and BCK Algebras", *Logique et Analyse* 28, 33–40. jstor.org/stable/44084112.

H. Barendregt, M. Bunder and W. Dekkers (1993), "Systems of Illative Combinatory Logic Complete for First-Order Propositional and Predicate Calculus", *The Journal of Symbolic Logic* 58, 769–788. doi: 10.2307/2275096.

D. J. Dougherty (1993), "Higher-Order Unification via Combinators", *Theoretical Computer Science* 114, 273–298. doi: 10.1016/0304-3975(93)90075-5.

I. Bethke, J. W. Klop and R. de Vrijer (1999), "Extending Partial Combinatory Algebras", *Mathematical Structures in Computer Science* 9, 483–505. doi: 10.1017/S0960129599002832.

Wolfram Guide Page (accessed March 26, 2021), "Combinatory Logic". reference.wolfram.com/language/guide/CombinatoryLogic.html.

Models of Combinatory Logic

G. D. Plotkin (1972), "A Set-Theoretical Definition of Application", Technical Report MiP-R-95, University of Edinburgh. doi: 10.1.1.62.4516.

C. Böhm and M. Dezani-Ciancaglini (1974), "Combinatorial Problems, Combinator Equations and Normal Forms", in *Automata, Languages and Programming*, J. Loeckx (ed.), International Colloquium on Automata, Languages, and Programming 1974, *Lecture Notes in Computer Science* vol. 14, Springer, 185–199. doi: 10.1007/978-3-662-21545-6_13.

G. Jacopini (1975), "A Condition for Identifying Two Elements of Whatever Model of Combinatory Logic", in *λ-Calculus and Computer Science Theory*, C. Böhm (ed.), International Symposium on Lambda-Calculus and Computer Science Theory 1975, *Lecture Notes in Computer Science* vol. 37, Springer, 213–219. doi: 10.1007/BFb0029527.

D. Scott (1975), "Combinators and Classes" in *λ-Calculus and Computer Science Theory*, C. Böhm, International Symposium on Lambda-Calculus and Computer Science Theory 1975, *Lecture Notes in Computer Science* vol. 37, Springer, 1–26. doi: 10.1007/BFb0029517.

D. M. R. Park (1976), "The Y-Combinator in Scott's Lambda-Calculus Models", University of Warwick, Department of Computer Science, Theory of Computation Report (unpublished). wrap.warwick.ac.uk/46310/1/WRAP_Park_cs-rr-013.pdf.

M. W. Bunder (1979), "Scott's Models and Illative Combinatory Logic", *Notre Dame Journal of Formal Logic* 20, 609–612. doi: 10.1305/ndjfl/1093882667.

E. Engeler (1981), "Algebras and Combinators", *Algebra Universalis* 13, 389–392. doi: 10.1007/BF02483849.

A. Meyer (1982), "What Is a Model of the Lambda Calculus?", *Information and Control* 52(1), 87–122. doi: 10.1016/S0019-9958(82)80087-9.

R. K. Meyer, M. W. Bunder and L. Powers (1991), "Implementing the 'Fool's Model' of Combinatory Logic", *Journal of Automated Reasoning* 7, 597–630. doi: 10.1007/BF01880331.

A. Asperti and A. Ciabattoni (1995), "On Completability of Partial Combinatory Algebras" in *Proceedings of the Italian Conference on Theoretical Computer Science*, A. de Santis (ed.), Italian Conference on Theoretical Computer Science, World Scientific, 162–175. logic.at/staff/agata/ictcs.pdf.

I. Bethke and J. W. Klop (1995), "Collapsing Partial Combinatory Algebras", in *Higher-Order Algebra, Logic, and Term Rewriting*, G. Dowek, et al. (eds.), International Workshop on Higher-Order Algebra, Logic, and Term Rewriting 1995, *Lecture Notes in Computer Science* vol. 1074, Springer, 57–73. doi: 10.1007/3-540-61254-8_19.

J. M. Dunn and R. K. Meyer (1997), "Combinators and Structurally Free Logic", *Logic Journal of the IGPL* 4, 505–537. doi:10.1093/jigpal/5.4.505.

Y. Akama (2004), "Limiting Partial Combinatory Algebras", *Theoretical Computer Science* 311, 199–220. doi: 10.1016/S0304-3975(03)00360-8.

K. Nour (2005), "Classic Combinatory Logic", *Computational Logic Applications,* Computational Logic and Applications 2005. arXiv:0905.1100.

P. Tarau (2015), "On a Uniform Representation of Combinators, Arithmetic, Lambda Terms and Types", in *Proceedings of the 17th International Symposium in Principles and Practice of Declarative Programming*, International Symposium in Principles and Practice of Declarative Programming 2015, Association for Computing Machinery, 244–255. doi: 10.1145/2790449.2790526.

Relations to Lambda Calculus

A. Church (1941), "The Calculi of Lambda-Conversion", *Annals of Mathematics Studies* 6, Princeton University Press. doi: 10.1515/9781400881932.

H. Barendregt (1970), "A Universal Generator for the Lambda Calculus", unpublished manuscript. wolframcloud.com/obj/combinators/Bibliography/Barendregt -AUniversalGeneratorForTheLambdaCalculus.

B. Robinet (1979), *Lambda Calcul et Sémantique Formelle des langages de programmation* (in French), Litp-Ensta.

G. Pottinger (1985), "Pleasant Beta-Conversion Using Combinators" (abstract), *The Journal of Symbolic Logic* 50, 1101–1102. doi: 10.2307/2274003.

H.-G. Oberhauser (1986), "On the Correspondence of Lambda Style Reduction and Combinator Style Reduction", in *Graph Reduction*, J. H. Fasel and R. M. Keller (eds.), Workshop on Graph Reduction 1986, *Lecture Notes in Computer Science* vol. 279, Springer, 1–25. doi: 10.1007/3-540-18420-1_47.

G. Huet (1989), *Initiation au λ-calcul* (in French), PN. apps.dtic.mil/dtic/tr/fulltext/u2/a218095.pdf.

S. Gilezan (1993), "A Note on Typed Combinators and Typed Lambda Terms", *Zbornik Radova Prirodno-Matematichkog Fakulteta Serija za Matematiku Review of Research Faculty of Science Mathematics Series* 23, 319–329. emis.de/journals/NSJOM/Papers/23_1/NSJOM_23_1_319_329.pdf.

M. W. Bunder (1996), "Lambda Terms Definable as Combinators", *Theoretical Computer Science* 169, 3–21. doi: 10.1016/S0304-3975(96)00111-9.

J. P. Seldin (2011), "The Search for a Reduction in Combinatory Logic Equivalent to λβ-Reduction", *Theoretical Computer Science* 412, 4905–4918. doi: 10.1016/j.tcs.2011.02.002.

H. Barendregt, W. Dekkers and R. Statman (2013), *Lambda Calculus with Types*, Perspectives in Logic, Cambridge University Press.

O. Kiselyov (2018), "λ to SKI, Semantically", in *Functional and Logic Programming*, J. Gallagher and M. Sulzmann (eds.), International Symposium on Functional and Logic Programming 2018, *Lecture Notes in Computer Science* vol. 10818, Springer, 33–50. doi: 10.1007/978-3-319-90686-7_3.

Relations to Type Theory

L. E. Sanchis (1964), "Types in Combinatory Logic", *Notre Dame Journal of Formal Logic* 5, 161–180. doi: 10.1305/ndjfl/1093957876.

R. Hindley (1969), "The Principal Type-Scheme of an Object in Combinatory Logic", *Transactions of the American Mathematical Society* 146, 29–60. doi: 10.2307/1995158.

J. P. Seldin (1969), "General Models for Type Theory Based on Combinatory Logic" (abstract), *Journal of Symbolic Logic* 34, 544. doi:10.2307/2270969.

J. Staples (1973), "Combinator Realizability of Constructive Finite Type Analysis", *Cambridge Summer School in Mathematical Logic*, A. R. D. Mathias and H. Rogers (eds.), *Lecture Notes in Mathematics* vol. 337, Springer, 253–273. doi: 10.1007/BFb0066777.

A. Rezus (1986), "Semantics of Constructive Type Theory", *Libertas Mathematica* 6, 1–82. system.lm-ns.org/index.php/lm/article/view/292.

K. Meinke (1991), "Equational Specification of Abstract Types and Combinators", in *Computer Science Logic*, E. Börger, et al. (eds.), International Workshop on Computer Science Logic 1991, *Lecture Notes in Computer Science* vol. 626, Springer, 257–271. doi: 10.1007/BFb0023772.

M. W. Bunder (1992), "Combinatory Logic and Lambda Calculus with Classical Types", *Logique & Analyse* 137/138, 69–79. jstor.org/stable/44084341.

J. R. Hindley (1997), *Basic Simple Type Theory*, Cambridge University Press. doi: 10.1017/CBO9780511608865.

S. Broda and L. Damas (2000), "On Principal Types of Combinators", *Theoretical Computer Science* 247, 277–290. doi: 10.1016/S0304-3975(99)00086-9.

J. Rehof and P. Urzyczyn (2011), "Finite Combinatory Logic with Intersection Types" in *Typed Lambda Calculi and Applications*, L. Ong (ed.), International Conference on Typed Lambda Calculi and Applications 2011, *Lecture Notes in Computer Science* vol. 6690, Springer, 169–183. doi: 10.1007/978-3-642-21691-6_15.

B. Düdder, et al. (2012), "Bounded Combinatory Logic", in *Computer Science Logic (CSL'12)—26th International Workshop/21st Annual Conference of the EACSL*, P. Cègielski and A. Durand (eds.), Computer Science Logic 2012, Leibniz-Zenturm fuer Informatik, 243–258. doi: 10.4230/LIPIcs.CSL.2012.243.

J. Bessai (2019), "A Type-Theoretic Framework for Software Component Synthesis", Dortmund University Library. doi: 10.17877/DE290R-20320.

Relations to Recursive Functions

H. B. Curry (1964), "Combinatory Recursive Objects of All Finite Types", *Bulletin of the American Mathematical Society* 70, 814–817. doi: 10.1090/S0002-9904-1964-11245-3.

W. W. Tait (1967), "Intensional Interpretations of Functionals of Finite Type I", *Journal of Symbolic Logic* 32, 198–212. doi: 10.2307/2271658.

G. Mitschke (1972), "Lambda-definierbare Funktionen auf Peanoalgebren" (in German), *Archiv für mathematische Logik und Grundlagenforschung* 15, 31–35. doi: 10.1007/BF02019773.

J. R. Hindley and G. Mitschke (1977), "Some Remarks about the Connections between Combinatory Logic and Axiomatic Recursion Theory", *Archiv für mathematische Logik und Grundlagenforschung* 18, 99–103. doi: 10.1007/BF02007262.

D. G. Skordev (1980), "Combinatory Spaces and Recursiveness in Them", Sofia.

J. A. Zashev (1983), "Recursion Theory in Partially Ordered Combinatory Models", dissertation, Sofia University.

J. A. Zashev (1987), "Recursion Theory in B-Combinatory Algebras", *Serdica* 12, 225–237. Zbl: 0654.03034.

N. Georgieva (1993), "Spaces with Combinators", *Archive for Mathematical Logic* 32, 321–339. doi: 10.1007/BF01409966.

M. Goldberg (2005), "On the Recursive Enumerability of Fixed-Point Combinators", *BRICS Report Series* 12. doi: 10.7146/brics.v12i1.21867.

Relations to Other Mathematical Structures

H. B. Curry (1934), "Foundations of the Theory of Abstract Sets from the Standpoint of Combinatory Logic" (abstract), *Bulletin of the American Mathematical Society* 40, 654. ams.org/journals/bull/1934-40-09/S0002-9904-1934-05918-4/S0002-9904-1934-05918-4.pdf.

A. Church (1937), "Combinatory Logic as a Semigroup" (abstract), *Bulletin of the American Mathematical Society* 43, 333. ams.org/journals/bull/1937-43-05/S0002-9904-1937-06526-8/S0002-9904-1937-06526-8.pdf.

E. J. Cogan (1955), "A Formalization of the Theory of Sets from the Point of View of Combinatory Logic", dissertation, University of Pennsylvania, *Zeitschrift für Mathematische Logik und Grundlagen der Mathematik* 1, 198–240. doi: 10.1002/malq.19550010304.

M. W. Bunder (1969), "Set Theory Based on Combinatory Logic", dissertation, University of Amsterdam.

T. F. Fox (1970), "Combinatory Logic and Cartesian Closed Categories" dissertation, McGill University. escholarship.mcgill.ca/concern/theses/6h440t871?locale=en.

C. Böhm (1971), "The Combinator Semigroup Has Two Generators", I. S. I., Turin, unpublished manuscript. wolframcloud.com/obj/combinators/Bibliography/Bohm -TheCombinatorSemigroupHasTwoGenerators.

M. W. Bunder (1974), "Propositional and Predicate Calculuses Based on Combinatory Logic", *Notre Dame Journal of Formal Logic* 15, 25–34. doi: 10.1305/ndjfl/1093891196.

M. W. Bunder (1974), "Various Systems of Set Theory Based on Combinatory Logic", *Notre Dame Journal of Formal Logic* 15, 192–206. doi: 10.1305/ndjfl/1093891298.

A. S. Kuzičev (1974) "Deductive Operators of Combinatory Logic", *Moscow University Mathematics Bulletin* 29, 8–14. wolframcloud.com/obj/combinators/Bibliography /Kuzicev-DeductiveOperatorsofCombinatoryLogic.

A. S. Kuzičev (1974), "On the Expressive Potentialities of Deductive Systems of λ-Conversion and Combinatory Logic", *Moscow University Mathematics Bulletin* 29, 58–64. wolframcloud.com/obj/combinators/Bibliography/Kuzicev-OnTheExpressive PotentialitiesofDeductiveSystemsofLambdaconversionandCombinatoryLogic.

J. Staples (1974), "Combinator Realizability of a Constructive Morse Set Theory", *Journal of Symbolic Logic* 39, 226–234. doi: 10.2307/2272635.

H. B. Curry (1975), "Representation of Markov Algorithms by Combinators" in *The Logical Enterprise*, A. R. Anderson, et al. (eds.), Yale University Press, 109–119.

M. Bel (1977), "An Intuitionistic Combinatory Theory Not Closed under the Rule of Choice", *Indagationes Mathematicae* 80, 69–72. doi: 10.1016/1385-7258(77)90032-4.

M. W. Bunder (1979), "On the Equivalence of Systems of Rules and Systems of Axioms in Illative Combinatory Logic", *Notre Dame Journal of Formal Logic* 20, 603–608. doi: 10.1305/ndjfl/1093882666.

M. R. Holmes (1991), "Systems of Combinatory Logic Related to Quine's 'New Foundations'", *Annals of Pure and Applied Logic* 53, 103–133. doi: 10.1016/0168-0072(91)90052-N.

Combinator Computation

Combinator Evaluation

T. J. W. Clarke, et al. (1980), "SKIM—The S, K, I Reduction Machine", in *LFP '80: Proceedings of the 1980 ACM Conference on LISP and Functional Programming*, Conference on LISP and Functional Programming 1980, Association for Computing Machinery, 128–135. doi: 10.1145/800087.802798.

N. D. Jones and S. S. Muchnick (1982), "A Fixed-Program Machine for Combinator Expression Evaluation", in *LFP '82: Proceedings of the 1982 ACM Symposium on LISP and Functional Programming*, D. P. Friedman and D. M. R. Park (eds.), Conference on LISP and Functional Programming 1982, Association for Computing Machinery, 11–20. doi: 10.1145/800068.802130.

D. R. Brownbridge (1985), "Cyclic Reference Counting for Combinator Machines", in *Functional Programming Languages and Computer Architecture*, J. P. Jouannaud (ed.), Conference on Functional Programming Languages and Computer Architecture 1985, *Lecture Notes in Computer Science* vol. 201, Springer 273–288. doi: 10.1007/3-540-15975-4_42.

C. L. Hankin, P. E. Osmon and M. J. Shute (1985), "Cobweb—A Combinator Reduction Architecture", in *Functional Programming Languages and Computer Architecture*, J. P. Jouannaud (ed.), Conference on Functional Programming Languages and Computer Architecture 1985, *Lecture Notes in Computer Science* vol. 201, Springer, 99–112. doi: 10.1007/3-540-15975-4_32.

K. Noshita and T. Hikita (1985), "The BC-Chain Method for Representing Combinators in Linear Space", *New Generation Computing* 3, 131–144. doi: 10.1007/BF03037065.

R. Milner (1985), "Parallel Combinator Reduction Machine", in *The Analysis of Concurrent Systems*, B. T. Denvir, et al. (eds.), *Lecture Notes in Computer Science* vol. 207, Springer, 121–127. doi: 10.1007/3-540-16047-7_41.

M. Takeichi (1985), "An Alternative Scheme for Evaluating Combinator Expressions", *Journal of Information Processing* 7, 246–253. id.nii.ac.jp/1001/00059885.

M. Scheevel (1986), "NORMA: A Graph Reduction Processor", in *Proceedings of the 1986 ACM Conference on LISP and Functional Programming*, Conference on LISP and Functional Programming 1986, Association for Computing Machinery, 212–219. doi: 10.1145/319838.319864.

M. Castan, M.-H. Durand and M. Lemaître (1987), "A Set of Combinators for Abstraction in Linear Space", *Information Processing Letters* 24, 183–188. doi: 10.1016/0020-0190(87)90183-9.

K. Noshita and X.-X. He (1987), "A Fast Algorithm for Translating Combinator Expressions with BC-Chains", *New Generation Computing* 5, 249–257. doi: 10.1007/BF03037465.

A. Piperno (1987), "A Compositive Abstraction Algorithm for Combinatory Logic", in *TAPSOFT '87*, H. Ehrig, et al. (eds.), International Joint Conference on Theory and Practice of Software Development 1987, *Lecture Notes in Computer Science* vol. 250, Springer, 39–51. doi: 10.1007/BFb0014971.

P. H. Hartel and A. H. Veen (1988), "Statistics on Graph Reduction of SASL Programs", *Software: Practice and Experience* 18, 239–253. doi: 10.1002/spe.4380180305.

A. Contessa, et al. (1989), "MaRS, a Combinator Graph Reduction Multiprocessor", *in PARLE '89 Parallel Architectures and Languages Europe*, E. Odijk, M. Rem and J. C. Syre (eds.), International Conference on Parallel Architectures and Languages Europe 1989, *Lecture Notes in Computer Science* vol. 365, Springer, 176–192. doi: 10.1007/3540512845_39.

P. J. Koopman and P. Lee (1989), "A Fresh Look at Combinator Graph Reduction (Or, Having a TIGRE by the Tail)", in *Proceedings of the SIGPLAN '89 Conference on Programming Language Design and Implementation*, Special Interest Group on Programming Languages 1989, ACM Press, 110–119. doi: 10.1145/74818.74828.

P. Koopman, P. Lee and D. P. Siewiorek (1990), "Cache Performance of Combinator Graph Reduction" in *Proceedings of the International Conference on Computer Languages 1990*, International Conference on Computer Languages 1990, IEEE, 39-48. doi: 10.1109/ICCL.1990.63759.

P. H. Hartel (1991), "Performance of Lazy Combinator Graph Reduction", *Software: Practice and Experience* 21, 299–329. doi: 10.1002/spe.4380210306.

M. A. Musicante and R. D. Lins (1991), "GM-C: A Graph Multi-combinator Machine", *Microprocessing and Microprogramming* 31, 81–84. doi: 10.1016/S0165-6074(08)80048-8.

M. Waite, B. Giddings and S. Lavington (1991), "Parallel Associative Combinator Evaluation", in *Parle '91 Parallel Architectures and Languages Europe*, E. H. L. Aarts, J. Leeuwen and M. Rem (eds.), *Lecture Notes in Computer Science* vol. 505, Springer, 771–788. doi: 10.1007/978-3-662-25209-3_47.

A. Dickinson and M. T. Pope (1992), "A Simple Machine (Based on the SK-Combinator Reduction Mechanism)", in *Proceedings from TENCON'92 - Technology Enabling Tomorrow*, IEEE Region 10 Conference 1992, IEEE, 845–849. doi: 10.1109/TENCON.1992.271851.

S. Thompson and R. Linus (1992), "The Categorical Multi-combinator Machine: CMCM", *The Computer Journal* 35, 170–176. doi: 10.1093/comjnl/35.2.170.

S. Kahrs (1993), "Compilation of Combinatory Reduction Systems", in *Higher-Order Algebra, Logic, and Term Rewriting*, J. Heering, et al. (eds.), International Workshop on Higher-Order Algebra, Logic, and Term Rewriting 1993, *Lecture Notes in Computer Science* vol. 816, Springer, 169–188. doi: 10.1007/3-540-58233-9_9.

S. D. Swierstra, P. R. A. Alcocer and J. Saraiva (1998), "Designing and Implementing Combinator Languages", in *Advanced Functional Programming*, S. D. Swierstra, J. N. Oliveira, and P. R. Henriques (eds.), International School on Advanced Functional Programming 1998, *Lecture Notes in Computer Science* vol. 1608, Springer, 150–206. doi: 10.1007/10704973_4.

jpt4 (2017), "A Relational SKI Combinator Calculus Interpreter", GitHub. github.com/jpt4/skio.

Wolfram Guide Page (accessed March 10, 2021), "Combinator Functions". wolframcloud.com/obj/wolframphysics/Tools/combinators.

Compilation to Combinators

S. L. Peyton Jones (1982), "An Investigation of the Relative Efficiencies of Combinators and Lambda Expressions", in *Proceedings of the 1982 ACM Symposium on LISP and Functional Programming*, Conference on LISP and Functional Programming 1982, Association for Computing Machinery, 150–158. doi: 10.1145/800068.802145.

F. W. Burton (1983), "A Linear Space Translation of Functional Programs to Turner Combinators", *Information Processing Letters* 14, 201–204. doi: 10.1016/0020-0190(82)90014-X.

T. Hikita (1984), "On the Average Size of Turner's Translation to Combinator Programs", *Journal of Information Processing* 7, 164–169. Zbl: 0386.68009.

S. Hirokawa (1985), "Complexity of the Combinator Reduction Machine", *Theoretical Computer Science* 41, 289–303. doi: 10.1016/0304-3975(85)90076-3.

M. S. Joy, V. J. Rayward-Smith and F. W. Burton (1985), "Efficient Combinator Code", *Computer Languages* 10, 211–224. doi: 10.1016/0096-0551(85)90017-7.

K. Noshita (1985), "Translation of Turner Combinators in O($n \log n$) Space", *Information Processing Letters* 20, 71–74. doi: 10.1016/0020-0190(85)90066-3.

R. S. Bird (1987), "A Formal Development of an Efficient Supercombinator Compiler", *Science of Computer Programming* 8, 113–137. doi: 10.1016/0167-6423(87)90017-7.

H. Nielson and F. Nielson (1990), "Functional Completeness of the Mixed λ-Calculus and Combinatory Logic", *Theoretical Computer Science* 70, 99–126. doi: 10.1016/0304-3975(90)90155-B.

M. S. Joy and V. J. Rayward-Smith (1995), "NP-Completeness of a Combinator Optimization Problem", *Notre Dame Journal of Formal Logic* 36, 319–335. doi: 10.1305/ndjfl/1040248462.

O. Kiselyov (1999, accessed April 15, 2021), "Many Faces of the Fixed-Point Combinator". okmij.org/ftp/Computation/fixed-point-combinators.html.

A. Hoogewijs and P. Audenaert (2003), "Combinatory Logic, a Bridge to Verified Programs", *Logic Colloquim*, Proceedings, Association of Symbolic Logic. hdl.handle.net/1854/LU-802976.

B. Lynn (2019, accessed April 15, 2021), "An Award-Winning Compiler". crypto.stanford.edu/~blynn/compiler/ioccc.html.

B. Lynn (accessed April 15, 2021), "LC to CL, Semantically". crypto.stanford.edu/~blynn/compiler/sem.html.

Combinators in Functional Programming

C. Böhm (1972), "Combinatory Foundation of Functional Programming", in *LFP '82: Proceedings of the 1982 ACM Symposium on LISP and Functional Programming*, Conference on LISP and Functional Programming 1982, Association for Computing Machinery, 29–36. doi: 10.1145/800068.802132.

S. K. Abdali (1974), "A Combinatory Logic Model of Programming Languages", thesis, University of Wisconsin. geomete.com/abdali/papers/phdthesis.pdf.

P. M. Maurer and A. E. Oldehoeft (1983), "The Use of Combinators in Translating a Purely Functional Language to Low-Level Data-Flow Graphs", *Computer Languages* 8, 27–45. doi: 10.1016/0096-0551(83)90004-8.

M. Wand (1983), "Loops in Combinator-Based Compilers", *Information and Control* 57, 148–164. doi: 10.1016/S0019-9958(83)80041-2.

P. Hudak and D. Kranz (1984), "A Combinator-Based Compiler for a Functional Language", *Proceedings of the 11th ACM SIGACT-SIGPLAN Symposium on Principles of Programming Languages*, Principles of Programming Languages, Association of Computing Machinery, 122–132. doi: 10.1145/800017.800523.

P.-L. Curien (1985), "Categorical Combinatory Logic", in *Automata, Languages and Programming*, W. Brauer (ed.), International Colloquium on Automata, Languages, and Programming 1985, *Lecture Notes in Computer Science* vol. 194, Springer, 130–139. doi: 10.1007/BFb0015738.

J. Gibert (1985), "The J-Machine: Functional Programming with Combinators", in *EUROCAL '85*, B. F. Caviness (ed.), European Conference on Computer Algebra 1985, *Lecture Notes in Computer Science* vol. 204, Springer, 197–198. doi: 10.1007/3-540-15984-3_262.

P. Hudak and B. Goldberg (1985), "Serial Combinators: 'Optimal' Grains of Parallelism", in *Functional Programming Languages and Computer Architecture*, J. P. Jouannaud (ed.), Conference on Functional Programming Languages and Computer Architecture 1985, *Lecture Notes in Computer Science* vol. 201, Springer, 382–399. doi: 10.1007/3-540-15975-4_49.

P. Bellot (1986), "Graal: A Functional Programming System with Uncurryfied Combinators and Its Reduction Machine", in *ESOP 86*, B. Robinet and R. Wilhelm (eds.), European Symposium on Programming 1986, *Lecture Notes in Computer Science* vol. 213, Springer, 82–98. doi: 10.1007/3-540-16442-1_6.

C. L. Hankin, G. L. Burn and S. L. Peyton Jones (1986), "A Safe Approach to Parallel Combinator Reduction (Extended Abstract)", in *ESOP 86*, B. Robinet and R. Wilhelm (eds.), European Symposium on Programming 1986, *Lecture Notes in Computer Science* vol. 213, Springer, 99–110. doi: 10.1007/3-540-16442-1_7.

I. Toyn and C. Runciman (1986), "Adapting Combinator and SECD Machines to Display Snapshots of Functional Computations", *New Generation Computing* 4, 339–363. doi: 10.1007/BF03037389.

A. Cheese (1987), "Combinatory Code and a Parallel Packet-Based Computational Model", *ACM SIGPLAN Notices* 22, 49–58. doi: 10.1145/24714.24720.

J. Gibert (1987), "Functional Programming with Combinators", *Journal of Symbolic Computation* 4, 269–293. doi: 10.1016/S0747-7171(87)80009-3.

S. R. D. Meira (1987), "Strict Combinators", *Information Processing Letters* 24, 255–258. doi: 10.1016/0020-0190(87)90144-X.

A. D. Robinson (1987), "The Illinois Functional Programming Interpreter", in *Papers of the Symposium on Interpreters and Interpretive Techniques*, SIGPLAN Symposium, Association for Computing Machinery, 64–73. doi: 10.1145/29650.29657.

C. Böhm (1988), "Functional Programming and Combinatory Algebras", in *Mathematical Foundations of Computer Science 1988*, M. P. Chytil, L. Janiga and V. Koubek (eds.), International Symposium on Mathematical Foundations of Computer Science 1988, *Lecture Notes in Computer Science* vol. 324, Springer, 14–26. doi: 10.1007/BFb0017128.

S. M. Sarwar, S. J. Hahn and J. A. Davis (1988), "Implementing Functional Languages on a Combinator-Based Reduction Machine", *ACM SIGPLAN Notices* 23, 65–70. doi: 10.1145/44326.44333.

P. P. Chu and J. A. Davis (1990), "Exploitation Fine-Grain Parallelism in a Combinator-Based Functional System", in *Third Symposium on the Frontiers of Massively Parallel Computation*, Third Symposium on the Frontiers of Massively Parallel Computation 1990, IEEE, 489–493. doi: 10.1109/FMPC.1990.89500.

D. Spencer (1990), "A Survey of Categorical Computation: Fixed Points, Partiality, Combinators, … Control?", Oregon Graduate Institute, Department of Computer Science and Engineering, Technical Report No. CS/E 90-017. doi: 10.6083/2514nk87q.

E. Cherlin (1991), "Pure Functions in APL and J", in *Proceedings of the International Conference on APL '91*, International Conference on APL 1991, North-Holland, 88–93. doi: 10.1145/114054.114065.

J. Gateley and B. F. Duba (1991), "Call-by-Value Combinatory Logic and the Lambda-Value Calculus", in *Mathematical Foundations of Programming Semantics*, S. Brookes, et al. (eds.), International Conference on Mathematical Foundations of Programming Semantics 1991, *Lecture Notes in Computer Science* vol. 598, Springer, 41–53. doi: 10.1007/3-540-55511-0_2.

P.-L. Curien (1993), *Categorical Combinators, Sequential Algorithms, and Functional Programming*, Springer Science & Business Media. doi: 10.1007/978-1-4612-0317-9.

S. M. Sarwar and J. A. Davis (1994), "New Families of Combinators for Efficient List Manipulation", *Journal of Systems and Software* 27, 137–146. doi: 10.1016/0164-1212(94)90027-2.

N. Raja and R. K. Shyamasundar (1995), "Combinatory Formulations of Concurrent Languages", in *Algorithms, Concurrency and Knowledge*, K. Kanchanasut and J. J. Lévy (eds.), Asian Computing Science Conference 1995, *Lecture Notes in Computer Science* vol. 1023, Springer, 156–170. doi: 10.1007/3-540-60688-2_42.

Y. Lafont (1997), "Interaction Combinators", *Information and Computation* 137, 69–101. doi: 10.1006/inco.1997.2643.

B. Meyer (1998), "The Component Combinator for Enterprise Applications", *Journal of Object-Oriented Programming* 10, 5–9. archive.eiffel.com/general/editorial/1998/combinator.html.

M. Goldberg (2005), "A Variadic Extension of Curry's Fixed-Point Combinator", in *Higher-Order Symbolic Computation* 18, 371–388. doi: 10.1007/s10990-005-4881-8.

P. D. Mosses (2007), "VDM Semantics of Programming Languages: Combinators and Monads", *Formal Methods and Hybrid Real-Time Systems*, C. B. Jones, Z. Liu and K. Woodcock (eds.), *Lecture Notes in Computer Science* vol. 4700, Springer, 483–503. doi: 10.1007/978-3-540-75221-9_23.

P. Weaver, et al. (2007), "Constructing Language Processors with Algebra Combinators", in *Constructing Language Processors with Algebra Combinators*, Generative Programming and Component Engineering 2007, Association for Computing Machinery, 155–164. doi: 10.1145/1289971.1289997.

N. A. Danielsson (2010), "Total Parser Combinators", in *Proceedings of the 15th ACM SIGPLAN International Conference on Functional Programming*, International Conference on Functional Programming, Association for Computing Machinery, 285–296. doi: 10.1145/1863543.1863585.

M. Bolingbroke, S. Peyton Jones and D. Vytiniotis (2011), "Termination Combinators Forever", in *Proceedings of the 4th ACM Symposium on Haskell*, Haskell 2011, Association for Computing Machinery, 23–34. doi: 10.1145/2034675.2034680.

O. Danvy and I. Zerny (2011), "Three Syntactic Theories for Combinatory Graph Reduction", in *Logic-Based Program Synthesis and Transformation*, M. Alpuente (ed.), International Symposium on Logic-Based Program Synthesis and Transformation 2010, *Lecture Notes in Computer Science* vol. 6564. Springer, 1–20. doi: 10.1007/978-3-642-20551-4_1.

D. Devriese and F. Piessens (2011), "Explicitly Recursive Grammar Combinators", in *Practical Aspects of Declarative Languages*, J. Launchbury and R. Rocha (eds.), International Symposium on Practical Aspects of Declarative Languages 2011, *Lecture Notes in Computer Science* vol. 6539, Springer, 84–98. doi: 10.1007/978-3-642-18378-2_9.

T. Schrijvers, et al. (2012), "Search Combinators", *Constraints* 18, 269–305. doi: 10.1007/s10601-012-9137-8.

P. Martins, J. P. Fernandes and J. Saraiva (2013), "A Combinator Language for Software Quality Reports", *International Journal of Computer and Communication Engineering* 2, 377–382. doi: 10.7763/IJCCE.2013.V2.208.

T. Ridge (2014), "Simple, Efficient, Sound and Complete Combinator Parsing for All Context-Free Grammars, Using an Oracle", in *Software Language Engineering*, B. Combemale, et al. (eds.), International Conference on Software Language Engineering 2014, *Lecture Notes in Computer Science* vol. 8706, Springer, 261–281. doi: 10.1007/978-3-319-11245-9_15.

M. S. New, et al. (2017), "Fair Enumeration Combinators", *Journal of Functional Programming* 27, e19. doi: 10.1017/S0956796817000107.

M. Sičák and J. Kollár (2017), "Supercombinator Driven Grammar Reconstruction", in *Proceedings of 2017 IEEE 14th International Scientific Conference on Informatics*, International Scientific Conference on Informatics, IEEE, 322-326. doi: 10.1109/INFORMATICS.2017.8327268.

A. Farrugia (2018), "Combinatory Logic: From Philosophy and Mathematics to Computer Science", Junior College Multi-disciplinary Conference: Research, Practice and Collaboration: Breaking Barriers: Annual Conference, 307–320. um.edu.mt/library/oar/handle/123456789/38118.

Metaprogramming with Combinators

L. Fegaras, T. Sheard and D. Stemple (1992), "Uniform Traversal Combinators: Definition, Use and Properties", in *Automated Deduction—CADE-11*, D. Kapur (ed.), International Conference on Automated Deduction 1992, *Lecture Notes in Computer Science* vol. 607, Springer, 148–162. doi: 10.1007/3-540-55602-8_162.

M. Fuchs (1997), "Evolving Combinators", in *Automated Deduction—CADE-14*, W. McCune, International Conference on Automated Deduction 1997, *Lecture Notes in Computer Science* vol. 1249, Springer, 416–430. doi: 10.1007/3-540-63104-6_42.

M. Fuchs, D. Fuchs and M. Fuchs (1997), "Solving Problems of Combinatory Logic with Genetic Programming" in *Genetic Programming 1997: Proceedings of the Second Annual Conference*, J. R. Koza, et al. (eds.), Genetic Programming 1997, Morgan Kaufmann, 102–110. cs.ucl.ac.uk/staff/W.Langdon/ftp/papers/gp1997/Fuchs_1997_spclGP.pdf.

A. Hamfelt and J. F. Nilsson (1998), "Inductive Synthesis of Logic Programs by Composition of Combinatory Program Schemes", in *Logic-Based Program Synthesis and Transformation*, P. Flener (ed.), International Workshop on Logic Programming Synthesis and Transformation 1998, *Lecture Notes in Computer Science* vol. 1559, Springer, 143–158. doi: 10.1007/3-540-48958-4_8.

P. S. di Fenizio (2000), "A Less Abstract Artificial Chemistry", in *Artificial Life VII: Proceedings of the Seventh International Conference on Artificial Life*, M. A. Bedau, et al. (eds.), International Conference on Artificial Life 2000, MIT Press, 49–53.

J. Tromp (2002), "Kolmogorov Complexity in Combinatory Logic". citeseerx.ist.psu.edu/viewdoc/download?doi=10.1.1.8.7359&rep=rep1&type=pdf.

M. Stay (2005), "Very Simple Chaitin Machines for Concrete AIT", *Fundamenta Informaticae* 68, 231–247. arXiv:cs/0508056.

M. Szudzik (2006), "An Elegant Pairing Function", Special NKS 2006 Conference. szudzik.com/ElegantPairing.pdf.

F. Briggs and M. O'Neill (2008), "Functional Genetic Programming and Exhaustive Program Search with Combinator Expressions", *International Journal of Knowledge-Based and Intelligent Engineering Systems* 12, 47–68. doi: 10.3233/KES-2008-12105.

P. Liang, M. I. Jordan and D. Klein (2010), "Learning Programs: A Hierarchical Bayesian Approach", in *ICML'10: Proceedings of the 27th International Conference on International Conference on Machine Learning,* International Conference on Machine Learning 2010, Omnipress, 639–646. doi: 10.5555/3104322.3104404.

P. Scott and J. Fleuriot (2012), "A Combinator Language for Theorem Discovery", in *Intelligent Computer Mathematics,* J. Jeuring, et al. (eds.), International Conference on Intelligent Computer Mathematics 2012, *Lecture Notes in Computer Science* vol. 7362, Springer, 371–385. doi: 10.1007/978-3-642-31374-5_25.

M. Bellia and M. E. Occhiuto (2013), "DNA Tiles, Wang Tiles and Combinators", *Fundamenta Informaticae* 133, 105–121. doi: 10.3233/FI-2014-1065.

L. R. Williams (2015), "Programs as Polypeptides". arXiv:1506.01573.

W. P. Worzel and D. MacLean (2015), "SKGP: The Way of the Combinator", in *Genetic Programming Theory and Practice XII*, M. Kotanchek, R. Riolo and W. Worzel (eds.), Springer, 53–57. doi: 10.1007/978-3-319-16030-6_4.

G. Gerules and C. Janikow (2016), "A Survey of Modularity in Genetic Programming", in *2016 IEEE Congress on Evolutionary Computation (CEC)*, IEEE, 5034–5043. doi: 10.1109/CEC.2016.7748328.

D. Xiao, J.-Y. Liao and X. Yuan (2018), "Improving the Universality and Learnability of Neural Programmer-Interpreters with Combinator Abstraction". arXiv:1802.02696.

Specific Programming Tasks

P. Koopman and R. Plasmeijer (1999), "Efficient Combinator Parsers", in *Implementation of Functional Languages,* C. Clack, T. Davie and K. Hammond (eds.), Symposium on Implementation and Application of Functional Languages 1998, *Lecture Notes in Computer Science* vol. 1595, Springer, 120–136. doi: 10.1007/3-540-48515-5_8.

M. Finger and W. Vasconcelos (2000), "Sharing Resource-Sensitive Knowledge Using Combinator Logics", in *Advances in Artificial Intelligence*, M. C. Monard and J. S. Sichman (eds.), Ibero-American Conference on Artificial Intelligence 2000, *Lecture Notes in Computer Science* vol. 1952, Springer, 196–206. doi: 10.1007/3-540-44399-1_21.

G. J. Pace (2000), "The Semantics of Verilog Using Transition System Combinators" in *Formal Methods in Computer-Aided Design*, W. A. Hunt and S. D. Johnson (eds.), International Conference on Formal Methods in Computer-Aided Design 2000, *Lecture Notes in Computer Science* vol. 1954, Springer, 442–459. doi: 10.1007/3-540-40922-X_25.

D. Leijen (2001), "Parsec, a Fast Combinator Parser". users.cecs.anu.edu.au/~Clem.Baker-Finch/parsec.pdf.

A. J. Kennedy (2004), "Pickler Combinators", *Journal of Functional Programming* 14, 727–739. doi: 10.1017/S0956796804005209.

I. D. Peake and S. Seefried (2004), "A Combinator Parser for Earley's Algorithm". citeseerx.ist.psu.edu/viewdoc/download?doi=10.1.1.504.5998&rep=rep1&type=pdf.

J. N. Foster, et al. (2007), "Combinators for Bidirectional Tree Transformations: A Linguistic Approach to the View-Update Problem", *ACM Transactions on Programming Languages and Systems* 29, 17–81. doi: 10.1145/1232420.1232424.

R. Lämmel (2007), "Scrap Your Boilerplate with XPath-Like Combinators", *ACM SIGNPLAN Notices* 42, 137–142. doi: 10.1145/1190215.1190240.

S. Mazanek and M. Minas (2007), "Graph Parser Combinators", in *Implementation and Application of Functional Languages*, O. Chitil, Z. Horváth and V. Zsók (eds.), Symposium on Implementation and Application of Functional Languages 2007, *Lecture Notes in Computer Science* vol. 5083, Springer, 1–18. doi: 10.1007/978-3-540-85373-2_1.

S. D. Swierstra (2009), "Combinator Parsing: A Short Tutorial", in *Language Engineering and Rigorous Software Development*, A. Bove, et al. (eds.), International LerNet ALFA Summer School on Language Engineering and Rigorous Software Development 2008, *Lecture Notes in Computer Science* vol. 5520, Springer, 252–300. doi: 10.1007/978-3-642-03153-3_6.

T. Neward (2012), "The Working Programmer–Building Combinators", *MSDN Magazine* 27. docs.microsoft.com/en-us/archive/msdn-magazine/2012/january/the-working-programmer-building-combinators.

H. Bergier (2020), "How Combinatory Logic Can Limit Computing Complexity", *EPJ Web Conference* 244. doi: 10.1051/epjconf/202024401009.

Extensions & Applications

Extensions of Combinators

J. T. Kearns (1969), "Combinatory Logic with Discriminators", *The Journal of Symbolic Logic* 34, 561–575. doi: 10.2307/2270850.

N. D. Belnap Jr. (1970), "Strict Polyadic Combinatory Logics", "Strict Polyadic Lambda Calculus", "Semantics of Strict Polyadicity", University of Pittsburgh, Faculty of Arts and Sciences, unpublished manuscripts.

N. D. Goodman (1972), "A Simplification of Combinatory Logic", *The Journal of Symbolic Logic* 37, 225–246. doi: 10.2307/2272970.

A. S. Kuzičev (1973), "Consistent Extensions of Pure Combinatory Logic" (in Russian; abstract in English), *Vestnik Moskovskogo Universiteta . Serija I . Matematika, Mehanika* 28, 76–81.

F. B. Fitch (1980), "A Consistent Combinatory Logic with an Inverse to Equality", *Journal of Symbolic Logic* 45, 529–543. doi: 10.2307/2273420.

R. K. Meyer, K. Bimbó and J. M. Dunn (1998), "Dual Combinators Bite the Dust" (abstract), *Bulletin of Symbolic Logic* 4, 463–464. doi:10.2307/420959.

T. Jech (1999), "Some Results on Combinators in the System TRC", *The Journal of Symbolic Logic* 64, 1811–1819. doi: 10.2307/2586813.

K. Bimbó (2003), "The Church–Rosser Property in Dual Combinatory Logic", *The Journal of Symbolic Logic* 68, 132–152. doi: 10.2178/jsl/1045861508.

K. Bimbó (2004), "Semantics for Dual and Symmetric Combinatory Calculi", *Journal of Philosophical Logic* 33, 125–153. jstor.org/stable/30226803.

Y. Akama (2006), "SN Combinators and Partial Combinatory Algebras", in *Rewriting Techniques and Applications,* International Conference on Rewriting Techniques and Applications 1998, *Lecture Notes in Computer Science* vol. 1379, Springer, 302–316. doi: 10.1007/BFb0052378.

A. Di Pierro, C. Hankin and H. Wiklicky (2006), "On Reversible Combinatory Logic", *Electronic Notes in Theoretical Computer Science* 135, 25–35. doi: 10.1016/j.entcs.2005.09.018.

J. H. Andrews (2007), "An Untyped Higher Order Logic with Y Combinator", *The Journal of Symbolic Logic* 72, 1385–1404. doi: 10.2178/jsl/1203350794.

P. Hancock (2008), "AMEN: The Last Word in Combinators (A Naperian Meditation)". citeseerx.ist.psu.edu/viewdoc/summary?doi=10.1.1.585.6287.

B. Jay and T. Given-Wilson (2011), "A Combinatory Account of Internal Structure", *Journal of Symbolic Logic* 76, 807–826. doi: 10.2178/jsl/1309952521.

Combinatory Grammars & Linguistics

P.-L. Curien (1986), *Categorical Combinators, Sequential Algorithms and Functional Programming*, Research Notes in *Theoretical Computer Science*, Pitman Publishing. doi: 10.1007/978-1-4612-0317-9.

R. D. Lins (1987), "On the Efficiency of Categorical Combinators as a Rewriting System", *Software: Practice and Experience* 17, 547–559. doi: 10.1002/spe.4380170807.

M. Steedman (1987), "Combinatory Grammars and Parasitic Gaps", *Natural Language & Linguistic Theory* 5, 403–439. doi: 10.1007/BF00134555.

M. Steedman (1988), "Combinators and Grammars", in *Categorical Grammars and Natural Language Structures*, E. Bach, R. T. Oehrle and D. Weeler (eds.), Springer, 417–442. doi: 10.1007/978-94-015-6878-4_15.

H. Yokouchi and T. Hikita (1988), "A Rewriting System for Categorical Combinators with Multiple Arguments", *SIAM Journal on Computing* 19, 78–97. doi: 10.1137/0219005.

J. F. Nilsson (1989), "A Case-Study in Knowledge Representation and Reasoning with Higher-Order Combinators", in *Scandinavian Conference on Artificial Intelligence 89: Proceedings of the SCAI '89*, H. Jaakkola and S. Linnainmaa (eds.), Scandinavian Conference on Artificial Intelligence 1989, IOS Press, 37–48.

P. Simons (1989), "Combinators and Categorical Grammar", *Notre Dame Journal of Formal Logic* 30, 241–261. doi: 10.1305/ndjfl/1093635081.

H. Yokouchi (1989), "Church-Rosser Theorem for a Rewriting System on Categorical Combinators", *Theoretical Computer Science* 65, 271–290. doi: 10.1016/0304-3975(89)90104-7.

J. Villadsen (1991), "Combinatory Categorial Grammar for Intensional Fragment of Natural Language", in *Scandinavian Conference on Artificial Intelligence—91*, B. H. Mayoh (ed.), Scandinavian Conference on Artificial Intelligence 1991, IOS Press, 328–339.

I. Biskri and J.-P. Desclés (1995), "Applicative and Combinatory Categorial Grammar (from Syntax to Functional Semantics)" in *Recent Advances in Natural Language Processing*, R. Mitkov and N. Nicolov (eds.), John Benjamins Publishing Company, 71–84. doi: 10.1075/cilt.136.08bis.

J.-P. Desclés (2004), "Combinatory Logic, Language, and Cognitive Representations," in *Alternative Logics. Do Sciences Need Them?*, P. Weingartner (ed.), Springer, 115–148. doi: 10.1007/978-3-662-05679-0_9.

I. Biskri (2005), "Applicative and Combinatory Categorial Grammar and Subordinate Constructions in French", *International Journal on Artificial Intelligence Tools* 14, 125–136. doi: 10.1142/S0218213005002028.

M. Steedman and J. Baldridge (2006), "Combinatory Categorial Grammar", in *Encyclopedia of Language & Linguistics (Second Edition)*, K. Brown (ed.), Elsevier, 610–621. doi: 10.1016/B0-08-044854-2/02028-9.

F. Hoyt and J. Baldridge (2008), "A Logical Basis for the D Combinator and Normal Form in CCG", in *Proceedings of ACL-08: HLT*, J. Allen, et al. (eds.), Annual Meeting of the Association for Computational Linguistics with the Human Language Technology Conference, Association for Computational Linguistics, 326–334. aclweb.org/anthology/P08-1038.pdf.

J. Kang and J.-P. Desclés (2008), "Korean Parsing Based on the Applicative Combinatory Categorial Grammar", in *Proceedings of the 22nd Pacific Asia Conference on Language, Information and Computation*, R. E. O. Roxas (ed.), Pacific Asia Conference on Language, Information and Computation, De La Salle University, 215–224. aclweb.org/anthology/Y08-1021.pdf.

C. Bozşahin (2012), "The Lexicon, Argumenthood and Combinators", in *Combinatory Linguistics*, De Gruyter, 31–42. library.oapen.org/bitstream/id/1eb35764-bbdd-444c-b2ad-13981a63672f/1005450.pdf.

C. Bozşahin (2012), "Syntacticizing the Combinators", in *Combinatory Linguistics*, De Gruyter, 43–60. library.oapen.org/bitstream/id/1eb35764-bbdd-444c-b2ad-13981a63672f/1005450.pdf.

C. Bozşahin (2013), *Combinatory Linguistics*, De Gruyter, doi: 10.1515/9783110296877.

M. Steedman (2018), "The Lost Combinator", *Computational Linguistics* 44, 613–620. doi: 10.1162/coli_a_00328.

S. T. Piantadosi (2021), "The Computational Origin of Representation", *Minds and Machines* 31, 1–58. doi: 10.1007/s11023-020-09540-9.

Confusing Issues

The term "combinatory analysis" has nothing to do with "combinators"; it's an earlier name for "combinatorics", used for example in:

P. A. MacMahon (1915), *Combinatory Analysis*, The University Press (Cambridge).

"Combinatory" is also not used in the sense of combinators in:

E. Post (1936), "Finite Combinatory Processes—Formulation 1", *The Journal of Symbolic Logic* 1, 103–105. doi: 10.2307/2269031.

"Combinator" is sometimes used as a fairly general term for a function or operation that combines computational operations, as in:

R. Milner (1982), "Four Combinators for Concurrency", in *Proceedings of the First ACM SIGACT-SIGOPS Symposium on Principles of Distributed Computing*, Symposium on Principles of Distributed Computing, Association for Computing Machinery, 104–110. doi: 10.1145/800220.806687.

L. Cardelli and R. Davies (1999), "Service Combinators for Web Computing", in *Transactions on Software Engineering*, IEEE, 309–316. doi: 10.1109/32.798321.

"The Combinator" is a recent combinatorial tool for idea generation, that seems to have no relation to combinators:

J. Han, et al. (2018), *The Combinator—A Computer-Based Tool for Creative Idea Generation Based on a Simulation Approach*, Cambridge University Press. doi: 10.1017/dsj.2018.7.

Y Combinator is a startup accelerator founded by P. Graham et al. in 2005; its name is derived from the fixed-point combinator, but otherwise it is unrelated:

Y Combinator (accessed March 22, 2021). www.ycombinator.com.

Thanks

Thanks to Henk Barendregt, Ariela Böhm, Emanuele Böhm, Michele Böhm, Martin Bunder, Mariangiola Dezani-Ciancaglini, Silvia Ghilezan, Roger Hindley, Oleg Kiselyov, Jan Willem Klop, Gerd Mitschke, Alberto Pettorossi and Adrian Rezus for suggestions and material for this bibliography, and to Paige Bremner, Amy Simpson and the University of Illinois library for extensive work in tracking down documents.

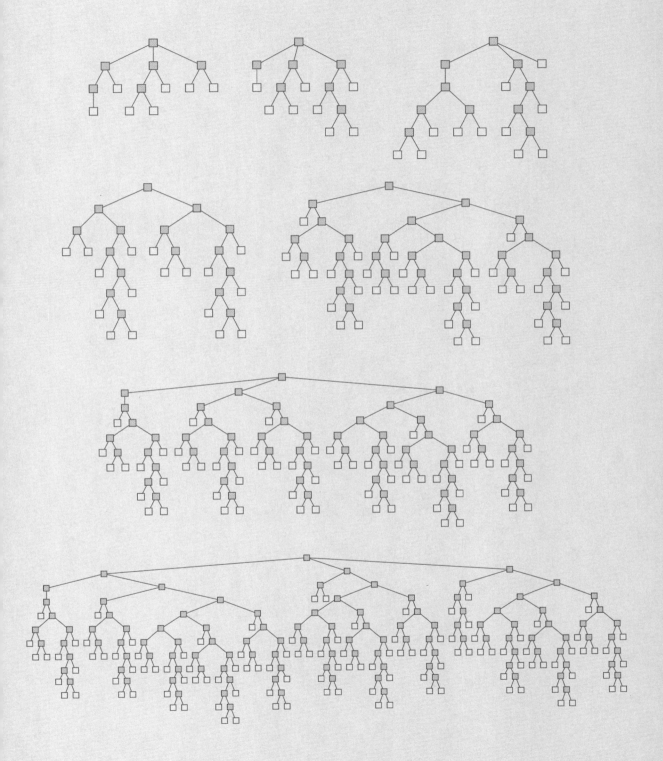

Index